Energy &
Resource
Development of
Continental
Margins

Pergamon Titles of Related Interest

Constans Marine Sources of Energy
Fazzolare and Smith Changing Energy Use Futures
Jones The Impact of North Sea Hydrocarbons
Karam and Morgan Energy and the Environment: Cost-Benefit Analysis
Ross Energy From Waves

Related Journals*

Computers and Geosciences
Deep Sea Research
Environment International
Environmental Professional
Geochimica et Cosmochimica Acta
Marine Pollution Bulletin
Ocean Engineering

*Specimen copies available upon request

Energy & Resource Development of Continental Margins

Edited by
Teh Fu Yen
Don Walsh

Pergamon Press

New York • Oxford • Toronto • Sydney • Frankfurt • Paris

Pergamon Press Offices:

U.S.A.	Pergamon Press Inc., Maxwell House, Fairview Park, Elmsford, New York 10523, U.S.A.
U.K.	Pergamon Press Ltd., Headington Hill Hall, Oxford OX3 0BW, England
CANADA	Pergamon of Canada, Ltd., Suite 104, 150 Consumers Road, Willowdale, Ontario M2J 1P9, Canada
AUSTRALIA	Pergamon Press (Aust.) Pty. Ltd., P.O. Box 544, Potts Point, NSW 2011, Australia
FRANCE	Pergamon Press SARL, 24 rue des Ecoles, 75240 Paris, Cedex 05, France
FEDERAL REPUBLIC OF GERMANY	Pergamon Press GmbH, Hammerweg 6, Postfach 1305, 6242 Kronberg/Taunus, Federal Republic of Germany

Library of Congress Cataloging in Publication Data

Symposium on Energy and Resource Development of
  Continental Margins, University of Southern
  California, 1979.
  Energy and resource development of continental
margins.

  Includes bibliographies and index.
  1. Petroleum in submerged lands—Congresses.
2. Gas, Natural, in submerged lands—Congresses.
3. Continental margins—Congresses. 4. Energy
development—Congresses. I. Yen, Teh Fu, 1927-
II. Walsh, Don, 1931-   III. Title.
TN863.S93   1979   333.8'23'09141   80-14813
ISBN 0-08-025127-7

Printed in the United States of America

CONTENTS

III. Environmental Constraints and Pollutant Analysis

IV. Social Impact and Safety of Onshore and Offshore Activities

PREFACE

Fossil Fuels are distributed worldwide and one of the potential sources is continental margins, many areas of which, however, are still unexplored.

In the coming decade, oil and gas potential are critically in need of exploration and development. With this in mind, the University of Southern California with the financial support from Finnigan Corporation sponsored the Symposium on Energy and Resource Development of Continental Margins at the University of Southern California from March 8-10, 1979.

Most participants of the Symposium felt the papers presented had significant importance and would have an impact on future related research. Consequently, it is meaningful to have them published to arouse the awarness of the public as well as the scientists engaged in related fields.

As a result, we have asked Pergamon Press to publish this book. Papers in this volume are selected from those presented at the Symposium. The cochairmen Drs. Teh Fu Yen and Don Walsh, served as editors. The contents of the book is divided into four sections: Geological and Geochemical Exploration, Novel Energy and Resource Recovery Methods, Environmental Constraints and Pollutants Analysis, and Social Impact and Safety of Onshore and Offshore Activities.

As editors, we are very happy to see this book published, and we would like to express our thanks to the speakers of the papers for making the manuscripts available, to Pergamon Press for their interest which made the publication of this work possible, to editing assistant, Sherry Hsi, and to all who helped with the typing and proofreading of the manuscripts.

Editors

LIST OF CONTRIBUTORS AND THEIR AFFILIATIONS

Numbers in parantheses indicate the pages on which the authors' contribution begin.

SUSAN H. ANDERSON (221), Sea Grant Program, University of Southern California, Los Angeles, California 90007

DONALD B. BRIGHT (216), The Port of Long Beach, California 90801

JOSEPH S. DEVINNY (224), Environmental Engineering Program, University of Southern California, Los Angeles, California 90007

R. E. FINNIGAN (114), Finnigan Corporation, Sunnyvale, California 94086

G. W. FODOR (142), Southwest Research Institute, San Antonio, Texas 78284

L. HALL (167), Environmental Resources Group, Los Angeles, California 90016

THOMAS L. HENYEY (10), Department of Geological Science, University of Southern California, Los Angeles, California 90007

M. T. JOHNSON (95), Department of Petroleum Enigneering, School of Earth Sciences, Stanford University, Stanford, California 94305

CRANDALL D. JONES (3), Exxon Company, Houston, Texas 77001

S. KAHANE (167), Environmental Resources Group, Los Angeles, California 90016

F. K. KAWAHARA (142), Environmental Monitoring and Support Laboratory, U. S. Environmental Protection Agency, Cincinnati, Ohio 45268

K. KIM (167), Pacific Environmental Service, Inc., Santa Monica, CA 90404

ROBERT B. KRUEGER (197), Law Offices of Nossaman, Krueger & Marsh, Los Angeles, California 90071

BERNARD LEMEHAUTE (41), Ocean Engineering, University of Miami, Miami, Florida 33149

S. S. MARSDEN (95), Department of Petroleum Engineering, School of Earth Sciences, Stanford University, Stanford, California 94305

F. M. NEWMAN (142), Southwest Research Institute, San Antonio, Texas 78284

MIKIHIKO OGURI (62), Harbors Environmental Projects, Institute for Marine and Coastal Studies, University of Southern California, Los Angeles, California 90007

LLOYD L. PHILIPSON (181), Institute of Safety and Systems Management, University of Southern California, Los Angeles, California 90007

DAVID W. SCHOLL (37), U.S. Geological Survey, Pacific-Arctic Branch of Marine Geology, Menlo Park, California 94025

CARLETON B. SCOTT (37), Union Oil Company of California, Union Oil Center, Los Angeles, California 90051

E. W. SCOTT (159), U.S. Geological Survey, Branch of Oil and Gas Resources, Laguna Niguel, California 92677

H. S. SILVUS, JR. (142), Southwest Research Institute, San Antonio, Texas 78284

DOROTHY F. SOULE (62), Harbors Environmental Projects, Institute for Marine and Coastal Studies, University of Southern California, Los Angeles, California 90007

JOHN D. SOULE (62), Harbors Environmental Project, Institute for Marine and Coastal Studies, University of Southern California, Los Angeles, California 90007

DALE STRAUGHAN (103), Institute for Marine and Coastal Studies, University of Southern California, Los Angeles, California 90007

JAMES I. S. TANG (52), Environmental Engineering Program, School of Engineering, University of Southern California, Los Angeles, California 90007

TEH FU YEN (52), Department of Chemical Engineering and Program of Environmental Engineering, University of Southern California, Los Angeles, California 90007

SECTION   I

<u>GEOLOGICAL AND GEOCHEMICAL EXPLORATION</u>

Exploring for Oil and Gas on the Continental Shelf

Seismic Risk and Resource Development in the Coastal Zone

Estimated Undiscovered Recoverable Oil and Gas Resources of
the Continental Shelf of the United States

Offshore Frontier Basins of the Pacific States and Alaska:
Some Generalizations about Their Geologic Setting and Petroleum Potential

1

EXPLORING FOR OIL AND GAS ON THE CONTINENTAL SHELF

by

Crandall D. Jones
Exxon Company, U.S.A.

Introduction

The practice by cartographers of vertical exaggeration has fostered the mistaken belief that the continent is surrounded by submarine cliffs. In actuality, there is a total of 853 thousand square miles of continental shelf surrounding the United States which slopes gently into the abysal deep. The vast majority of it presents no obstacles in drilling for oil and gas. Except by drilling, there is no way to confidently measure the reserves of an area. Offshore drilling normally requires a number of successful wells in an area before the endeavor can be considered economically feasible. But, because of the lengthy time lag between discovery and production, exploration of our remaining frontier areas should be rapidly pursed.

The Continental Shelf

The continental shelf is usually defined as that portion of the continent which extends from the shoreline seaward to a depth of approximately 200 meters (600 feet). The continental slope is the portion of the continent lying between ocean depths of 600 and 6,000 feet. That portion of the North American continental shelf, bordered by the United States, comprises some 853,300 square miles, with the continental slope providing an additional 450,000 square miles. The continental shelf varies in width from area to area. For example, in the Gulf of Mexico and in the Bering Sea, there are broad continental shelves which extend more than a hundred miles from shore before the water depth exceeds 600 feet. In other areas, notably the West Coast, the continental shelf is much narrower. However, of the combined total of 1,300,000 square miles which constitute the area of the continental shelf and slope adjacent to the borders of the United States, only 3% has been leased for exploration for gas and oil to date.

Because maps attempting to depict the earth's surface in three dimensions must resort to considerable vertical exaggeration, many people have the mistaken idea that the continent is surrounded by precipitous submarine cliffs. Cartographers depart from the use of true scale in many maps, for instance, some of the maps published by the National Geographic Society. Comparison of these maps and true scale cross-sections often reveal that the vertical exaggeration of the former can be misleading.

There are three areas in the U.S. in which the outer-continental shelves have been explored fruitfully. They are the Gulf of Mexico (the most productive dis-

4

covery to date), Southern California, and the Upper Cook Inlet of Alaska. Additionally, some exploration has taken place along the coasts of Northern California, Oregon and Washington, and in the Gulf of Alaska. Exploration of areas off the Atlantic Coast and off much of Alaska has only just commenced.

Formation of Reservoirs

Oil and gas are complex hydrocarbon compounds formed by the alteration of organic matter which has accumulated and been buried in finegrained sediments deposited on the sea floor. This sediment-organic matter "sandwich" is subjected to pressure from the continued deposition of sediments in the basin, and to heat from the earth over a period of millions of years. Heat and pressure, along with bacterial action, transform the organic matter into oil and gas. Gradually, over a long period of time, the gaseous or liquid hydrocarbons migrate from the source beds into more porous and permeable sedimentary rocks which act as reservoirs. If favorable trapping mechanisms exist, further migration through the reservoir rock is prevented, causing the oil and gas to form a captive pool. These so-called "pools" are not caverns filled with oil, nor are they underground lakes or streams. The oil or gas is found between the grains of sand or limestone that make up the reservoir rock much as water is held by a kitchen sponge.

Figure 1 is a two-dimensional cross-section through a series of arched alternating beds of sand and shale. Geologists call this an anticline. The beds of sand are porous and permeable, while the beds of shale are impermeable. The gas, oil, and water have separated naturally, according to their relative densities. Gas, being the lightest, goes to the top, followed by oil, then water. In the lower sandstone bed, there is no oil, only gas. Sometimes oil occurs without the presence of free natural gas. All too frequently, the trap is filled only with water.

A TYPICAL TRAP

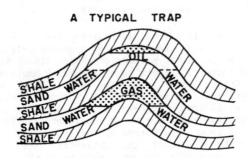

Fig. 1.

There are a number of different types of traps (Fig. 2). Sometimes the beds of alternating sand and shale have been displaced along a fault, which forms an impermeable barrier and, thus, traps the oil. Salt domes which flow up from an underlying thick bed of salt sometimes cause numerous reservoir traps to form around the periphery of the dome. Then, there is the typical stratigraphic trap, formed by the tilting and erosion of a series of layers, and the subsequent deposition of further layers. When the lowermost of the overlying layers is an impermeable shale, a trap is formed.

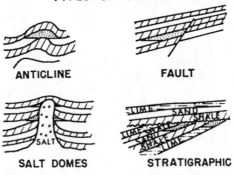

Fig. 2.

The four examples discussed are merely a few of many types of traps which exist.

Location of Reservoirs

In petroleum exploration, offshore geology is frequently inferred from a projection of onshore conditions and occurrence. The presence or absence of reservoir beds, as well as the presence or absence of hydrocarbon source beds can sometimes be successfully projected, but it is difficult to project the presence of traps. Magnetic surveys are sometimes carried out, using an airborne magnetometer. The purpose of this type of survey is to estimate the amount of sedimentary rocks which may be present, and to determine their depth. Oil and gas are usually generated and trapped in sedimentary rocks which are the sandstones, shales, and limestone deposited in the ocean. A gravity meter is ocassionally used, as well. It measures the pull of the earth's gravity, which varies from place to place, depending on the density of the rocks. Thus, the resulting gravity map, which gives an indication of the structural grain of the ocean floor, can be useful in detecting the presence of salt domes, since the salt is less dense than the beds of rock surrounding it. In some areas, scuba divers and small, manned submarines have examined the sea floor to determine the attitude of dip of the beds which outcrop and to obtain samples of these beds for age-determination.

Sometimes, core and stratigraphic test holes are punched or drilled into the ocean floor to determine age and rock type. Recovered samples are frequently analyzed for the presence of hydrocarbons to ascertain whether or not beds are present which could be sources of oil and gas. Deep holes are sometimes drilled in areas where traps are not believed to exist. These so-called "strat tests" are undertaken by groups of companies to acquire information which can be used by geochemists to predict the absence or presence of hydrocarbons in the sedimentary basin. If a basin becomes productive and is trapped by many wells, the subsurface information obtained can be used to project the geologies of nearby undrilled areas, lending greater confidence in the prediction of traps.

By far the most often used device for location of traps is the reflection seismograph. Sea-going vessels are commonly equipped with this equipment which emit sharp bursts of energy which travel through the water to penetrate various layers of the Earth until they are reflected and return to be picked up by a series of sensitive receivers affixed to cable which is towed by the vessel. Knowing the precise time the sound sources are emitted, the velocity of the sound waves through water and various layers of sedimentary rock, as well as the precise time the reflected signals are received by the geophone, it is possible to calculate the depth to the reflecting horizon. In the early days of marine seismic exploration, dynamite was used as a sound source. Today, the sound source is provided by the sudden release of compressed air or by the explosion of a mixture of propane and air within a rubber sleeve. The explosion causes the sleeve to expand and contract quickly, providing the proper sound source without danger to ocean life.

It costs about $400 per mile to make a seismic survey offshore, and a similar amount to process the raw data. It is not uncommon for a company to spend 5 or 6 million dollars acquiring and processing seismic information gathered in a frontier area.

At present, there is no technology available for the direct detection of hydrocarbons in the subsurface, except drilling. Seismic technology has, under certain ideal conditions, led to confidence in statements about the probable occurrence of natural gas and oil in the subsurface. These methods, however, fail to give a 100% guarantee of certainty. It is embarrassingly easy to make incorrect inferences about what exists below the surface. This is especially true in frontier areas where oil or gas production has not yet been proven. In such areas, assessments of trap occurrence is usually successful, however, less success is met with in forecasting the presence of reservoir rock. Assessments of hydrocarbon generation are frequently wrong, as well. Some basins are dry simply because no oil or gas was ever generated there, or because it was generated, but migrated and was lost to the ocean floor before any traps were formed.

Oil and gas exploration, while fascinating, can also be extremely disappointing.

Exploration Rights

Since most of the oil and gas in our continental shelves belongs to the federal government, the right to exploration requires strict adherence to the Outer Continental Shelf (OCS) leasing regulations. The process of acquiring lease rights involves a great number of steps which make the procedure a lengthy one. Following is a chronological list of those steps with the lengths of time usually needed for completion.
1) Call for Nominations
   When OCS issues the call, oil and gas companies nominate specific acreages to be put up for lease. By that time, the industry has usually acquired some reconnaisance-type geophysical information -- possibly a magnetic or gravity survey and often some seismic information. This allows the explorers to be somewhat selective in nominating tracts. In addition to these nominations, other interests make negative nominations, recommending that certain acreages not be made available for oil and gas exploration.
2) Deadline of Nominations
   Two months after the call for nominations has been made, the nominations are due.
3) Announcement of Tracts
   The nominations are forwarded to the Department of the Interior which, after perhaps 3 months of consideration, makes an announcement of the tracts tentatively to be offered. After this, industrial exploration goes into high gear. It is usually following the announcement that very detailed seismographic information is obtained, and shallow- and deep-core testing is made.

4) Draft of Environmental Statement
   Some 6 to 19 months after the tentative announcement, a draft Environmental
   Impact Statement is issued.
5) Public Hearing (2 months later).
6) Final Environmental Statement (3 to 4 months later).
7) Proposed Notice of Shale
   This notice follows the final environmental statement by 2 months, and gives
   some tentative details concerning the type of leases that will be issued,
   stipulations to be imposed, types of bidding arrangements, and more precise
   definition of the tracts.
8) Notice of Sale
   In another 2 months, the final notice is announced, carrying even more definite
   information on the number of tracts, lease stipulations, and the date and
   location of the sale. As well, it communicates decisions pertaining to who
   will and won't be allowed to bid, who can and can't bid together, and what the
   various bidding methods will be.
9) Sale
   A month later, the sale is held, presuming that there are no delays due to
   legal suits. By this time, the industrial concerns have acquired and interpreted their exploratory data. They have constructed geological maps covering
   the sale tracts. In areas of potential traps they have constructed economic
   models, forecasting the number of platforms to be erected and the number of
   wells which would have to be drilled if oil or gas are present. Future costs,
   prices, rates of production, and regulatory systems have been projected, along
   with the possibility of delays through lawsuits and problems arising in the
   procedures of acquiring permits. After this careful calculation, an assessment is made of the degree of competition, before a final bed is submitted
   for an offshore tract.
From 21 to 37 months will have elapsed between the original call for nominations
and the final sale.

Lease Royalties

An OCS lease presently contains 2,304 hectacres, or about 3,693 acres, which is
a plot of ocean floor with approximately 4.8 kilometers (3 miles) on a side. A
lease gives the bidder the exclusive right to explore for oil and gas on this
piece of ocean floor over a period of five years. Each year, the leasee is
charged a $3.00 per acre rental. If production has not been established by the
end of the 5-year term, the lease must be surrendered to the government. If production is established within that time, royalties of no less that 12.5% of the
oil or gas derived from the leased acreage must be paid to the government.

The vast majority of leases sold to date require a 1/6, or 16.67%, royalty. This
means that the government gets 1/6 of the oil or gas produced by the oil company.
The government can turn around and sell the oil and gas they receive in payment,
or can allow the oil company to sell it and return the proceeds from the sale.

Most lease sales to date have used the so-called "bonus bid" system, wherein the
royalty is fixed, usually at 1/6, and the bid variable is an initial cash bonus.
During the last few years, some leases have been sold at royalty rates greater
than 1/6. In the sale of South Atlantic leases recently, a number of the leases
were scheduled to be sold on the basis of a cash bonus, but with a sliding-scale
royalty which is a function of the value and rate of production.

Drilling Methods

Before the purchases of a lease has occurred, the oil company will have discussed the drilling methods to be used in each particular area.

Exploratory drilling can be done from an above-water platform upon legs temporarily attached to the ocean floor. These rigs are usually used in water depths ranging from a few feet to about 300 feet and are known as "jackup" rigs, companies must frequently drill in greater depths, other mobile rigs are used such as the drill-ship and the semisubmersible rig.

The drillship is a ship-shaped vessel specifically designed for drilling. A hole constructed through the center of the ship serves as the drilling point of the well. This type of vessel is frequently utilized for drilling in areas located a great distance from the base of operations, since it has a great capacity for storing the materials needed for lengthy periods of time.

The semisubmersible rig is generally used in areas of rough water, particularly the North Sea and the Gulf of Alaska. Its features are a stable, floating plat-form, usually anchored to the bottom with anywhere from 8 to 12 anchors.

Economic Consideration

The cost of operating an offshore rig varies considerably, depending on the type, age, size, water depth capabilities, etc. of the individual rig. However, total operating costs are likely to run from $20,000 to $130,000 a day. Since it may take anywhere from 10 to 200 days to drill a well, costs for individual wells can range from $200,000 to $26,000,000.

A basic difference between onshore and offshore oil and gas operations is cost-recovery. Onshore, if a wildcat discovers oil or gas, the incremental costs needed to place the well on actual production are usually very small. As a consequence, practically any well which is able to produce oil or gas and pay its daily oper-ating costs is immediately placed on production. The offshore, however, is a different matter. Platforms necessary to support production equipment and wells can cost anywhere from $10 million to $200 million. The result of such heavy investments is that it usually takes more than one successful well to "discover" an offshore field. It must be demonstrated that there is enough producible oil or gas in the trap to warrant any future investments necessary to place it on production. Therefore, a number of wells are usually drilled on an offshore prospect before a decision is made to install a platform.

Phases of Offshore Development

Figure 3 represents the various phases of offshore exploration and development. It will be recalled that the sale is held at least 2 years after the initial exploration activities commenced. Exploration activities build to peak during the exploratory drilling phase about 5 years from the initial date and taper off as the various prospects are thoroughly explored. After exploratory drilling commences, some discoveries are made and development drilling then commences. The high activity is accounted for by the number of people, and amount of equipment and material which must be transported during this phase.

Later, as development wells are completed, production commences and, in this ins-tance, is shown reaching a peak some 10 years after initial exploration commenced.

Production finally comes to an end some 30 years after exploration began. As other lease sales are held, the three phases -- exploration, development, and production -- will be carried on for the most recently acquired acreages. Thus, in the Gulf of Mexico, activity has been going on for about 40 years. Sizeable fields are still being discovered there, showing why it is virtually impossible in most basins to accurately assess the ultimate oil and gas potential after having drilled only a few wells.

Suggestion for Future Production

In the attempt to establish offshore oil and gas production over the last 30 years, over 22 thousand wells have been drilled, and over 8 billion barrels of oil and 45 trillion cubic feet of natural gas have been produced. Economically, through 1976, the government had received $18 billion in bonuses, $4.5 billion in royalties, and $180 million in rentals, totaling over $22 billion in revenue to the government from offshore activity. At the same time, the total value of the production to date has been $27.6 million: 83% of which has gone to the government, and 17% to industry.

The most recent OCS lease sales schedule shows a concentration of activity centering in the Gulf/Atlantic areas. In effect, emphasis is being shifted to activity in the already productive areas, while the relatively untapped Pacific coastal regions are given secondary consideration. Initial exploration should proceed as rapidly as possible in these frontier and virgin basins, since, as it has been demonstrated, it is uncertain which frontier basin will become the major oil and gas producers. Considering the time lag between initial discovery and peak production, exploration should be pursued with all possible dispatch.

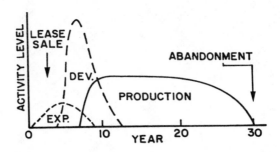

Fig. 3

2

SEISMIC RISK AND RESOURCE DEVELOPMENT
IN THE COASTAL ZONE

by

Thomas L. Henyey
University of Southern California

Introductory Statement

The identification of seismic hazards connected with onshore and offshore resource
development and facility sitings constitutes a primary task along many parts of
the U.S. coastal zone. The risk ranges from nominal along the eastern seaboard to
severe along the southern coast of Alaska. Countless projects have been, and many
more will be, strongly affected by limitations imposed as a result of the apparent
seismic risk. The economic and social impacts will be significant. The identification of seismic hazards in the coastal zone is uniquely challenging to the
seismologist and structural engineer. The data is sparse and it is difficult and
expensive to acquire; generally indirect methods of data gathering are required.
Finally, the surface geology often consists of very recent, poorly lithified materials, whose spatial extent and physical characteristics under seismic shaking are
poorly known.

The term "seismic risk" can be considered to apply to any or all of the following:

1. strong ground motion (10^{-1} to > g; 1 g = nominal value of the acceleration of
   gravity at the earth's surface).

2. weak ground motion (10^{-3} to 10^{-1} g).

3. faulting (primary or "tectonic" ground rupture and aseismic slip or creep).

4. secondary phenomena such as landsliding, liquefaction, tsunamis, etc.

Ground Motion

Earthquakes, whether local or distant, are complex phenomena that can impose severe
loadings on structures. They release large amounts of strain energy from limited
volumes of the earth's crust in the form of seismic body and surface waves.
Structures or facilities located near earthquake epicenters may be within the source
region, which is rebounding elastically to a condition of lower stress. According
to the concepts of the elastic rebound theory (Reid, 1911; Knopoff, 1958; Chinnery,
1961), horizontal and/or vertical elastic relaxation occurs within the source region.
However, although shaking in the source region is largely an elastic phenomenon, the
proximity to a finite dimensional source, coupled with an earthquake mechanism whose
rupture dimensions, time, history and sense of motion are imperfectly known, significantly complicates our ability to theoretically predict the nature of these
motions (e.g. Bolt, 1970).

The linear theory of elastic wave propagation can be used, in principle, to predict
the nature of ground motions outside the source region. However, severe constraints

12

are imposed by a general lack of knowledge of the source mechanism, transfer function (elastic and anelastic parameters along the seismic wave travel paths), and
the site function (the elastic and anelastic parameters of the near surface geology
at a particular site).

For specific structures at specific sites, the seismic information which can be
predicted, irrespective of distance from the source, is severely deficient.   Furthermore, no single parameter completely reflects the potential effects of an earthquake. Acceleration, velocity, displacement, direction of motion, frequency content, and duration of shaking all play important roles in determining effects on
soils, foundations, and superstructures (Page 1975; Cornell and Vanmarcke, 1975;
Bea, 1975). These parameters are, in general, strongly dependent upon the site
geology, regional geology and fault-site configuration. To date, engineers have
relied largely on a few empirically determined strong-motion accelerograms (i.e.
the time-history of earthquake acceleration) to provide insight into the forces
induced by strong earthquake shaking. Generally statistical prediction techniques
based on "standardized" or idealized accelerograms are used (e.g. see W.A.S.H. 1254).
Also "peak accelerations" (figure 1) are useful for qualitative analyses of the
severity of ground motions (For a more complete discussion of peak acceleration, see
Trifunac and Brady, 1976).

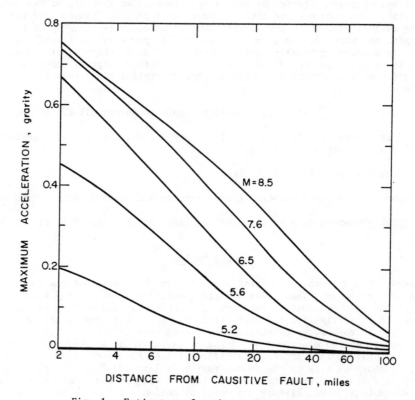

Fig. 1. Estimates of peak acceleration on rock as a
function of distance from causitive fault.
Results for different magnitude earthquakes
are shown. See Schnabel and Seed (1973).

Unfortunately, the existing strong-motion data suffer from a number of deficiencies pertinent to design criteria in the coastal zone. These are:

1. No strong-motion accelerograms exist to date for offshore sites, and very few for important onshore coastal sites.

2. Many of the standardized accelerogram response spectra on which statistical predictions are based, may not be appropriate for certain coastal zone structures, such as offshore platforms. In particular, the role of the water column is not well understood.

3. Offshore seismic sources may have important characteristics different from the onshore sources which have been used as a basis for much of our seismic knowledge.

4. The influence of water-saturated coastal zone sediments on seismic response in structures is poorly known.

5. Instabilities arising from long-duration "weak ground motion" which may be important in poorly consolidated coastal zone soils are virtually unknown.

Faults

The location and identification of active faults provides data on both the hazard from rupture and the loci of potential earthquakes. Assessing a given coastal zone site's potential hazard from fault rupture or aseismic slip (creep, subsidence) is a deceptively difficult task (CDMG Note #49). First, most faults are complex --- evidence for identifying an "active" fault versus a long inactive or dormant fault may be difficult to acquire. This distinction is complicated by the fact that dormant faults may be reactivated by underground pumping or fluid injection, either in the form of regional strain release (Pakser et al., 1969), or subsidence (Kovach, 1974; Castle and Yerkes, 1969; Barrows, 1974). Surface faulting parallel to pre-existing faults, and having vertical and horizontal offset, has been documented in areas of differential subsidence. Faults can also serve as conduits to the surface or between oil sands and aquifers for high pressure subsurface fluids.

Jennings (1973) has suggested criteria for recognizing active faults. For the purposes of delineating faults in the coastal zone, it is convenient to apply Jennings' criteria with some modification as follows:

1. those faults along which historically recorded large earthquakes or aseismic slip has occured.

    a. has a historic record of important earthquake activity but with no apparent surface rupture (e.g., 1933 Long Beach earthquake and Newport-Inglewood fault).

    b. has a historic record of at least one major earthquake with surface rupture.

    c. has a historic record of aseismic slip (creep).

2. those faults which have been active during the Quaternary (∿ last 2 million years).

    a. active during the Holocene (∿ last 11,000 years) as evidenced by offset of recent sediments where sedimentation has been essentially continuous. Often a clear distinction between Pleistocene and Holocene sediments can be made on the outer continental shelf (OCS).

b. active during the Pleistocene or significant portion thereof. Either the most recent sediments are not offset or this evidence is lacking.

3. those faults demonstrating alignments of micro-earthquake epicenters (M < 3) along the fault trace, particularly where evidence for surface rupture during the Quaternary is equivocal or lacking.

The evaluation of a given coastal zone site with respect to the potential hazard of fault rupture is based extensively on the concepts of recency and recurrence of slip. In a general way, the more recent the faulting, the greater the probability for near-future rupture (Ziony, et al., 1973). Also, where the accepted seismic risk is greater, a more liberal interpretation of "recency" may be used. For example, recurrence intervals between destructive events on the order of 100 years might be acceptable for a simple family residence, while a figure of 500,000 years might be deemed necessary for a nuclear reactor.

The terms "long inactive" or "dormant" are preferred to the term "inactive" when describing faults. This distinction is particularly significant within tectonically active regions of the earth such as the Cordillera of western North America, where faults may show a continuous spectrum of slip recurrence intervals from several years to thousands or even millions of years.

Faults, whether "active" or "dormant" according to the preceding scheme, pose a special set of potential seismic risks in the presence of oil field operations; these risks may be accentuated in the coastal zone:

1. Subsidence bowls develop in regions of subsurface fluid withdrawal; bowl formation is often associated with local faults (Kovach, 1974; Castle and Yerkes, 1969) along which a seismic slip and/or microearthquakes can occur. The severity of subsidence is inversely related to the degree of reservoir lithification.

2. Fluid repressuring during secondary oil recovery can cause creep or microearthquake related slip at depth along otherwise "locked" faults by reducing the normal stress across a fault plane under regional shear (Pakiser, et al., 1969; Teng, et al., 1973).

3. The locus of preferred regional strain energy release may be transferred from an active fault to a neighboring dormant fault in the presence of increased fluid pore pressure.

4. Fault zones can act as underground passage ways for fluid under pressure. "Short circuiting" of underground high-pressure reservoirs to fault zone during drilling can lead to undesirable migration of fluids along these fault planes.

Some remarks regarding the foregoing risks are in order. First, while (1) and (2) above have been well documented, (3) and (4) have not. However, although (4) is not an established risk, the correlation of seepage in the Santa Barbara Channel of southern California, with mapped faults (Fischer and Berry, 1973) attests to this hazard. Second, appreciable stored strain energy is not introduced by man into the subsurface inasmuch as the energy budget of his operations is negligible in comparison to energy released by potentially hazardous (M > 3) earthquakes. Thus, it is concluded that such operations can at best "trigger" strain energy release. However, it is important to remark that subsurface casing, drill rod, and pipelines can fail under only a few centimeters of dislocation across a fault plane. Third, Teng and Henyey (1974) have suggested that sedimentary basins may be largely stress-decoupled from the underlying basement. Thus oil field operations in "soft-sediments" may be less likely to trigger stress-release than those which involve penetration into "hard-rock." Fourth, the existence of intra-plate stresses acting across faults in relatively aseismic areas; i.e., away from present plate boundaries (Sykes and Sbar, 1973; also the New Madrid and Charleston earthquakes) does not permit ruling out the triggering of seismic events in such areas.

Secondary Risks

Secondary seismic risks result from strong ground motion in the epicentral region. Important secondary risks are as follows:

1. Submarine landsliding - the downslope release of unconsolidated material along basin peripheries, continental slopes, submarine canyons, etc., in the form of slumps, liquified soil flow, or turbidity currents.

2. Liquefaction - the liquifying of water-saturated sands, silts, and/or muds by shaking, thereby reducing or eliminating the support provided by the seafloor for platforms, pipelines, etc.

3. Tsunami - seismic sea wave generated by vertical movements of the seafloor during faulting or by submarine landsliding. Tsunami effects are most severe at the sea-coast interface, presenting hazards to low-lying, waterfront struc- tures, nearshore terminal and docking facilities and moored vessels during off- loading and onloading. The possibility of tsunami damage to offshore platforms and/or drilling vessels cannot be discounted. The critical factors to be con- sidered are the height and velocity of the resultant sea waves at the site in question, and the particular stage of drilling of development of a well at the time a tsunami strikes (Kovach, 1974).

Seismicity

Of fundamental importance to hazard assessment is the local and regional seismicity. Characterization of the seismicity included two key elements:

1. the spatial and temporal distribution of earthquakes with respect to magnitude.

2. the magnitudes of the maximum credible and maximum probable earthquakes (CDMG Note #43), their recurrence intervals, and their probable epicentral locations.

The maximum credible earthquake is the maximum earthquake that appears capable of occurring under the presently known tectonic framework. It is a believable event that is in accord with all known geological and seismological facts. In determining the maximum credible earthquake, little regard is given to its probability of oc- currence, except that its liklihood of occurring is great enough to be of concern. The following are considered when determining the maximum credible earthquake:

1. the seismic history of the region in question.

2. the length of the significant fault(s); as a rule of thumb, maximum surface rupture can be assumed to be approximately ½ fault length (see figs. 2 & 3).

3. the type of faulting; for example, where statistics are available it has been shown that surface ruptures on reverse faults are shorter than on strikeslip faults for earthquakes of equivalent magnitude.

4. the regional tectonic setting.

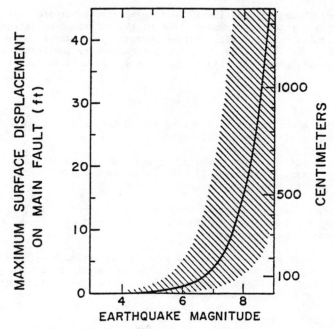

Fig. 2. Estimate of maximum surface displacement along fault as a
function of earthquake magnitude. Hatchered region gives
the range of observations. See Bonilla and Buchanan (1970);
also Lamar, et al. (1973).

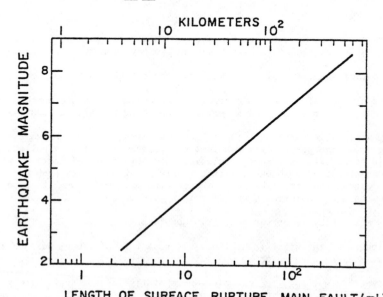

Fig. 3. Estimate for the length of surface rupture along a fault as
a function of earthquake magnitude. See Bonilla and Buchanan
(1970); also Wallace (1970).

The maximum probable earthquake is the maximum earthquake that is likely to occur over a specified interval of time (e.g. projected facility's lifetime), and might be considered the "functional-basis" earthquake for seismic design criteria. Inasmuch as statistics on these events are generally poor, if not non-existent, the maximum probable earthquake and its recurrence interval are generally estimated from the statistics on smaller events. Local and regional networks are used to provide frequency versus magnitude plots which generally fit the analytical relationship:

$$N = 10^{a-bM}$$

where N is the number of events of magnitude M and larger per unit area unit time, and a and b are constants characteristic of the seismo-tectonic province (Richter, 1958). The constant b, or "b-value" provides, by extrapolation, estimates on the recurrence interval for larger events not adequately represented in the data set (Allen, et al., 1965; Henyey and Teng, 1976). Clearly, extrapolation past the maximum credible earthquake would not be made.

Seismicity studies also generally involve fault-plane solutions when possible. That is, the directions of slip and orientations of the fault planes are determined from seismogram first motion data for representative seismic events with adequate network coverage. These data not only provide information on fault "grain" but also yield the direction of maximum compressive stress.

Coastal Zone Seismicity in Southern California

General patterns

The historical interest and importance of seismicity in southern California, coupled with this region's extensive development of its coastal zone, provides an important case history in seismic risk.

Station control for earthquake epicenters has been provided largely by the California Institute of Technology, with significantly improved coverage in the last 10 years from the U.S. Geological Survey, California Division of Mines and Geology, and the University of Southern California (e.g., figure 4). However, significant gaps in station coverage still exist. These are largely in the offshore area, around Points Conception and Arguello and between Long Beach and San Diego. The most serious of these gaps is in the offshore area particularly in view of potential offshore development.

An accurate delineation of seismicity (that is, spatial and temporal distributions, frequency-magnitude statistics) is dependent on several important factors: (1) station coverage, (2) network detectability limits, (3) duration of monitoring, and (4) earthquake frequency. Earthquake frequency and duration of monitoring are interrelated, and become increasingly more important for small events, which are most useful for delineating faults and defining regional seismicity. Along the coastal zone, events smaller than magnitude 1.5 are essentially undetectable. As the number of stations increase, the number of small events detected also increases, and thus an evaluation of station control (total number and locations) is essential for the prediction of temporal and spatial seismic relationships based on small magnitude events (microseismicity).

Earthquake epicenters in the coastal zone of southern California are plotted in Figures 5 and 6. Figure 5 covers the time period between 1932 and 1967, and Figure 6 covers the time period between 1968 and 1974. The subdivision into two time periods is convenient due to the improved station coverage in the coastal zone

18

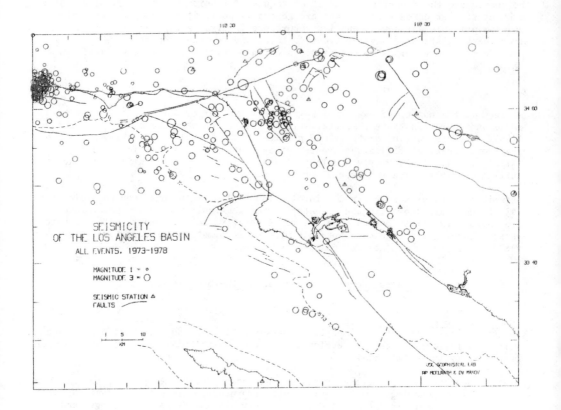

SEISMICITY
OF THE LOS ANGELES BASIN
ALL EVENTS. 1973-1978

MAGNITUDE 1 = ∘
MAGNITUDE 3 = ○

SEISMIC STATION △
FAULTS ⌒

5   10
KM

Fig. 4. Seismicity of the Los Angeles Basin between
1973 and 1978. Size of circle gives relative
magnitude.

beginning about 1968. Thus Figure 5 contains somewhat more precise epicenter locations.

The general level of seismicity in the coastal zone is similar to that of greater
southern California. What appears to be a decrease in seismicity, oceanward, may
in part be real, but also reflects the lack of station coverage on the outer continental shelf (continental borderland). Based on the most recent structural
studies in the continental borderland, seismicity should probably decrease to the
west since major faults in the outermost borderland are not as prevalent as in the
innermost borderland.

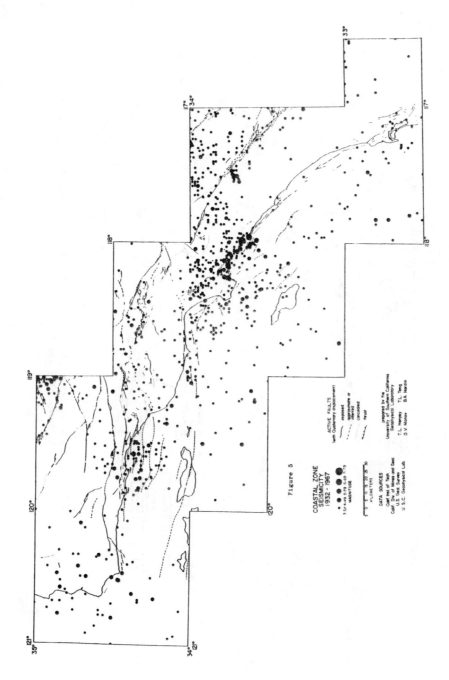

Fig. 5. Seismicity of southern California coastal zone between 1932 and 1967.

20

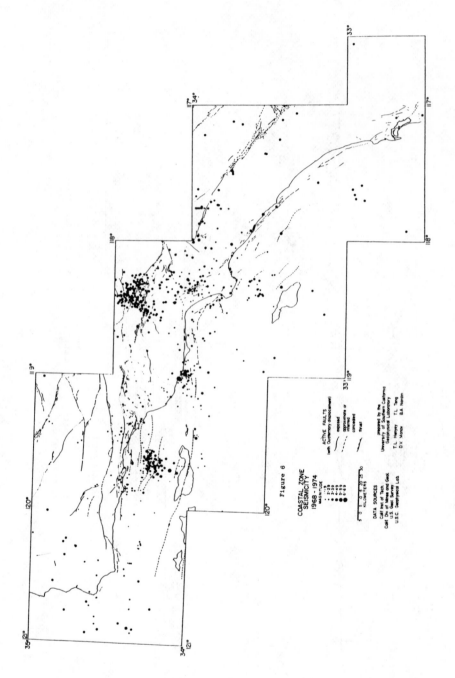

Fig. 6. Seismicity of southern California coastal zone between 1968 and 1974.

In addition to the apparent westward decrease in seismicity, Figures 5 and 6 show higher than average seismicity in the Los Angeles and eastern Santa Barbara basins, and in the vicinity of Points Arguello and Conception.

The apparent average seismicity of the Los Angeles Basin is, in part, but not wholly, due to the existence of better seismic station coverage in this region. The high seismicity in the eastern Santa Barbara Basin is real, and contrasts sharply with the western portion of the basin. Noteworthy is the presence of numerous east-west trending faults in the eastern part of the basin.

The apparent scatter of events near Point Conception can probably be attributed in large part to the lack of adequate station coverage here. Better station coverage may provide a correlation of events with important faults such as the Santa Ynez or San Simeon-Hosgri Fault systems (Hall, 1975).

Seismicity of the coastal zone can also be depicted using strain-release maps according to the method of Allen et al., 1965 (see Figures 7 and 8). To construct these maps, the coastal zone was broken into 5' by 5' quadrangles and events in a given quadrangle assigned 50% of their weight to that quadrangle and the remaining 50% to the surrounding quadrangles. Strain release was expressed in equivalent numbers of magnitude 3.0 earthquakes. Figure 7 is the result after two iterations of smoothing; Figure 8 is the result after 10 iterations. Clearly, principal foci of strain release are strongly correlated with the coastal zone. This is not an unexpected result in that most of the coastline from Point Arguello to San Diego is fault controlled.

Frequency-Magnitude Statistics

Figures 9, 10, and 11 are frequency vs. magnitude plots for earthquakes in the Los Angeles basin and Santa Barbara region. The logarithm of the cumulative number of seismic events (starting with the largest) is plotted versus magnitude or more specifically magnitude interval. An interval of ¼ of a Richter magnitude unit has been used.

The straight line (recurrence curve)

$$N = 10^{a-bM}$$

has been fit to the intermediate magnitude events for which the statistics are relatively complete; that is, they have not been biased appreciably by detectability limits and/or insufficient monitoring time. Nuttli (1974) discusses tests for statistically assessing the sufficiency of such a data base to compute recurrence curves. These tests have been applied to the data represented in figures 9 to 11 and indicate that the data is reasonably complete in the magnitude range $M = 3$ to $M = 5$, and can be used to generate the recurrence curves shown.

Figure 9 is a cumulative frequency vs. magnitude plot for the Los Angeles Basin-- here taken to be that region ($\sim$ 10,000 sq. km) between longitudes $117^\circ 45'W$ and $119^\circ 00'W$, and latitudes $33^\circ 15'N$ and $34^\circ 00'N$ (Figure 12). Figure 9 suggests that it is reasonable to expect one magnitude 6 to 6.5 earthquake or larger to recur once every 50 years within this region. The "b" value (from Equation 1) of 0.98 for the Los Angeles basin is similar to that reported by Allen, et al., 1965, for a common but larger area. It is somewhat higher than "b" values for other regions of southern California. High "b" values appear to be characteristic of crust under tension or low horizontal compressive stress and/or increased plasticity (Thatcher and Brune, 1971).

22

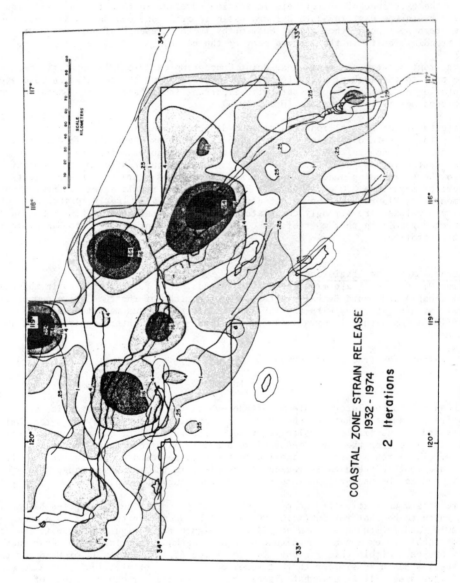

COASTAL ZONE STRAIN RELEASE
1932 - 1974

2 Iterations

Fig. 7. Strain release in the coastal zone of southern California
between 1932 and 1974; two iterations of smoothing.

23

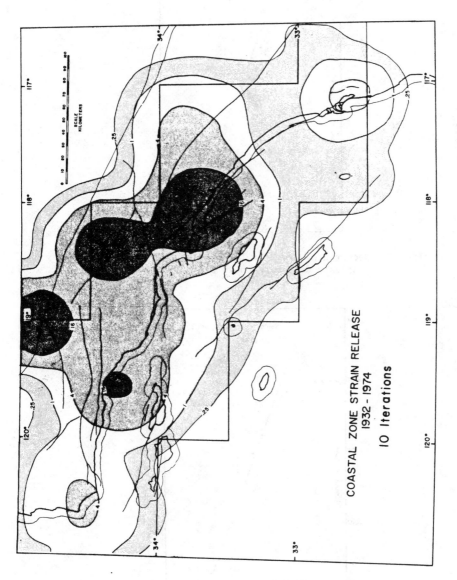

COASTAL ZONE STRAIN RELEASE
1932 - 1974
10 Iterations

Fig. 8. Strain release in the coastal zone of southern California between 1932 and 1974; then iterations of smoothing.

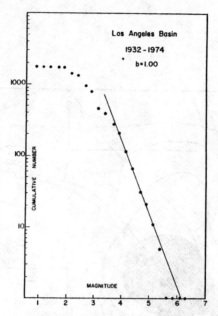

Fig. 9.   Recurrence curve for Los Angeles Basin from all
available data between 1932 and 1974.

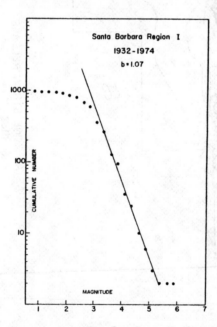

Fig. 10.   Recurrence curve for Santa Barbara region from
all available data between 1932 and 1974.

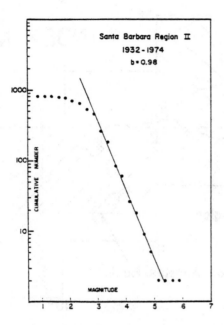

Fig. 11. Recurrence curve for Santa Barbara region from
selected data between 1932 and 1974.

Figures 10 and 11 are two recurrence curves for the Santa Barbara region (Figure 12), taken to be that region (∿ 20,000 sq. km) between longitudes 119°W, and also south of the San Andreas fault. Figure 10 includes all data between 1932 and 1974 and yields a "b" value of 1.07. The best fit suggests a recurrence of one magnitude 5.5 to 6.0 or greater approximately every 50 years somewhere in the region. It is important to note that this recurrence estimate may be too low by as much as a factor of two due to: (a) poor station control in the region between 1932 and the present, and (b) the occurrence in western Santa Barbara County of two additional earthquakes exceeding magnitude 6.0 in 1925 and 1927 not included in our data set. The recurrence interval of earthquakes exceeding magnitude 6 to 6.5, per unit area, for the Santa Barbara region is essentially the same as for the Los Angeles Basin. Figure 11 (Santa Barbara region II) includes the same data as Figure 10 less 3 major events in 1941, 1968, and 1973 which may bias the data. The 1941 and 1973 events were deleted due to incomplete data sets. The 1968 earthquake was deleted because it was an obvious swarm (several events of equivalent magnitude) rather than the usual main shock with characteristic aftershock sequence. The resultant "b" value is 0.98 or the same as for the Los Angeles Basin. Thus the tendency toward higher "b" values or swarm seismic behavior does not appear to be typical for this region.

Seismic Monitoring In The Southern California Coastal Zone

The geophysical laboratory at the University of Southern California has been conducting seismicity studies in coastal zone oil fields for the past several years. The principal objectives of this program are:

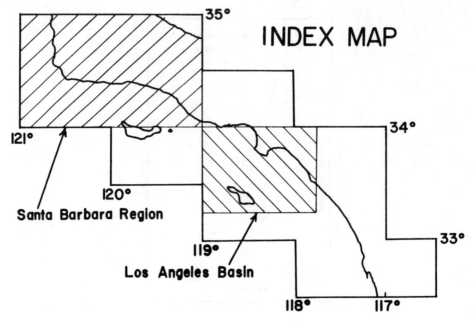

Fig. 12. Index map of southern California coastal zone.

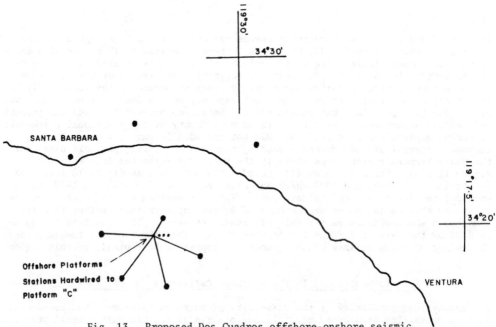

Fig. 13. Proposed Dos Cuadros offshore-onshore seismic
network. Solid circles indicate station locations.

1. to determine microearthquake activity and detect fault movement that might re-
    sult from fluid injection operations.

2. to accurately locate all earthquakes in the coastal zone for the purpose of
    identifying those events which might be man-induced from those which are
    naturally-occuring along the many active faults in southern California.

Two oil fields have been under study for the past six years. These include the
Wilmington oil field (Wilmington-Long Beach) and the Baldwin Hills oil field (West
Los Angeles). A third oil field - the Dos Cuadras offshore field, Santa Barbara
Channel - was instrumented by the end of 1978.

The Wilmington and Baldwin Hills fields are each monitored by a 5 element small-
aperture (local) array, including one instrument emplaced in a deep-well, together
with a large aperture peripheral (regional) network. All stations are land-based.
Station locations and epicenter locations from the last six years are shown in
figure

The Dos Cuadras field will be instrumented with an eight element array having 10-15
km aperture. Five stations will be placed on the sea floor and three on land along
the nearby coast (Fig. 13). An eight station regional network (including two
stations on the Channel Islands) operated by the U.S.G.S. and California Institute
of Technology will aid in epicenter control.

The potential interrelationships between man's sub-surface oil field activities and
southern California natural seismicity provides the impetus for this ongoing program
in coastal zone hazards research. The oil-rich Newport-Inglewood Uplift, slicing
through metropolitan Los Angeles county, has been the locus of a number of destructive earthquakes. These include the 1933 M6.3 Long Beach earthquake (120 lives
lost and 50 million dollars in property damage), the 1941 M5.0 Compton earthquake,
and the 1944 M5.5 Torrance earthquake. Earthquakes the size of the 1933 event can
be expected in the near future along the Newport-Inglewood fault, as well as along
other major coastal zone faults such as the Santa Monica Mountains-Malibu fault
zone and faults of the eastern Santa Barbara basin, which themselves have been the
sites of major historical events. The drastic increase in population and industry
within the coastal strip between Santa Barbara and San Diego since 1933 raises the
spectre of multibillion dollar losses for M6-7 type earthquakes.

The causal relationship between subsurface fluid pressure and earthquake occurrence
has long been suspected (Hubbert and Rubey, 1959) and more recently confirmed
(Evans, 1966; Hollister and Weimer, 1968; Pakiser, et al., 1969). Underground fluid
injection can be expected to increase pore pressures and thereby decrease the
effective normal stress and resistance to shearing, according to the following
relation:

$$\sigma_s = C + \mu(\sigma_n - P)$$

where σ_s = shearing resistance, σ_n = normal stress, P = pore pressure, μ = coefficient of friction, C = intrinsic constant

High pore pressure can thus be thought of as having a lubricating effect on faults in
the sense that it reduces frictional resistance to movement, but it does not have
a lubricating effect in the sense that μ itself is changed by pore pressure. In
fact, μ has been shown to be reasonably constant over a wide range of pore pressures
(Handin, et al., 1963). Reducing underground fluid pore pressures can be expected
to have the inverse of the effect described above. During the producing lifetime
of an oil field, pore pressures are initially reduced during the primary oil
extraction phase, while later, during secondary recovery techniques employing under-

ground fluid injection, the pore pressures are increased. Thus changes in the level and character of the local seismicity can be expected.

Summary

Earthquakes are a well established hazard within much of the U.S. coastal zone. However, large gaps exist in our knowledge of those factors directly responsible for seismic risk. For example no strong motion accelerograph data exists for free-field offshore sites.

Due to the presence of recent sediments and rapid rates of erosion, the location and identification of active faults in the coastal zone is difficult, often requiring indirect methods of investigation. In many places such as California and Alaska, coastline physiography is controlled by major active faults. Faults which may not be seismically active, also pose risks in the presence of underground fluids at elevated pressures.

Coastal zone seismic networks are necessary for delineating local and regional seismicity, which includes (a) the spatial and temporal distribution of earthquakes as a function of magnitude, (b) the magnitudes of the maximum credible and maximum probable earthquakes, and (c) the focal mechanism. Data on the recurrence intervals of "functional-basis" earthquakes for seismic design criteria are generally not available, but must be estimated from the statistics of smaller events. In regions such as southern California, where the seismic station coverage is good, the "functional-basis" earthquakes can be determined with a relatively high level of confidence.

The potential for man-induced seismicity resulting from subsurface oil-field operations is being investigated by U.S.C. in the seismically active southern California coastal zone. Instruments have been monitoring the Wilmington and Baldwin Hills oil fields in the Los Angeles basin for several years, and are presently being installed around the Dos Cuadros offshore field in Santa Barbara Channel. At present, the data from mature oil fields such as Wilmington and Baldwin Hills, where secondary recovery techniques are well established, suggest that hypocenters are largely confined to the crystalline basement rocks below the "soft" sedimentary oil producing zones. This may not be the case, however, in fields where the producing zones and basement are not differentiated by marked changes in lithology.

References

1. Allen, C.R., P. St. Amand, C.F. Richter, and J.M. Nordquist (1965), Relationship between seismicity and geologic structure in the southern California region. Bull. Seismic Soc. Am., 55, 753.

2. Barrows, A.G. (1974), A review of the geology and earthquake history of the Newport-Inglewood structural zone, southern California. California Division of Mines and Geology, Special Report 114, 115 p.

3. Bea, R.G. (1976), Earthquake criteria for platforms in the Gulf of Alaska. OTC Preprints, OTC 2675.

4. Bolt, B.A., (1970), Causes of earthquakes: in Earthquake Engineering, R.L. Wiegel, editor, Prentice-Hall, p. 21-45.

5. Bonilla, M.G. and J.M. Buchanan (1970), Interim report on worldwide historic surface faulting; U.S. Geol. Survey Open-File Report, 32 p.

6. Castle, R.O. and R.F. Yerkes (1969), Recent surface movements in the Baldwin Hills, Los Angeles County, California. A study of surface deformation associated with oil-field operations. U.S. Geol. Surv. Open File Report, 300 p.

7. CDMG Note #43 (1975), Guidelines for determining the maximum credible and the maximum probable earthquakes. California Division of Mines and Geology.

8. CDMG Note #49 (1975), Guidelines for evaluating the hazard of surface fault rupture. California Division of Mines and Geology.

9. Chinnery, M.A. (1961), The deformation of the ground around surface faults, Bull. Seismic Soc. Am., 51, 355.

10. Cornell, C.A. and E.H. vanMarcke (1975), Seismic risk analysis for offshore structures. OTC Preprints, OTC 2350.

11. Evans, D.M. (1966), The Denver area earthquakes and the Rocky Mountain disposal well. The Mountain Geologist, 3 (1), p. 23-26.

12. Fischer, P.J. and R. Berry (1973), Environmental hazards of the Santa Barbara Channel: Oil and gas seeps and Holocene faulting, in Geology, Seismicity, and Environmental Impact. Assoc. Eng. Geologists, Spec. Pub.

13. Hall, C.A., Jr. (1975), San Simeon-Hosgri fault system, coastal California: Economic and environmental implications. Science, 190, 1291.

14. Handin, J., R.V. Hager, Jr., M. Friedman, and J.N. Feather (1963), Experimental deformation of sedimentary rocks under confining pressure: pore-pressure tests. Am. Assoc. Pet. Geol. Bull., 47 (5) p. 718-755.

15. Henyey, T.L. and T.L. Teng (1976), Oil and tar seep studies on the shelves off southern California, III. Seismicity of the southern California coastal zone. U.S.C. Geophysical Laboratory Technical Report 75-4.

16. Hollister, J.C. and R.J. Weimer (1968), Geophysical and geological studies of the relationship between the Denver earthquakes and the Rocky Mountain arsenal well, Part A, Quarterly of the Colorado School of Mines, 63.

17. Hubbert, M.K., and W.W. Rubey (1959), Role of fluid pressure in mechanics of overthrust faulting. Geol. Soc. Amer. Bull., 70, p. 115-206.

18. Jennings, C.W. (1973), State of California preliminary fault and geologic map. California Division of Mines and Geology, Preliminary Report 13.

19. Knopoff, L. (1958), Energy release in earthquakes, Geophys. Jour., 1, 44-52.

20. Kovach, R.L. (1974), Source mechanisms for Wilmington oil field, California, Subsidence earthquakes, Bull. Seismic Soc. Am., 64, 699.

21. Lamar, D.L., P.M. Merifield, and R.J. Proctor (1973), Earthquake recurrence intervals on major faults in southern California in Geology, Seismicity, and Environmental Impact; Assoc. Eng. Geol. Spec. Pub., p. 265.

22. Nuttli, O.W. (1974), Magnitude-recurrence relation for central Mississippi Valley earthquakes. Bull. Seis. Soc. Amer., 64, p. 1189-1208.

23. Page, R.A. (1975), Evaluation of seismicity and earthquake shaking at offshore sites. OTC Preprints, OTC 2354.

24. Pakiser, L.C., J.P. Eaton, J.H. Healy, and C.B. Raleigh (1966), Earthquake prediction and control. Science, 166, 1467.

25. Reid, H.F. (1911), The elastic rebound theory of earthquakes, Bull. Dept. Geology, Univ. California, 6, 413.

26. Schnabel, P.B. and H.B. Seed (1973), Acceleration in rock for earthquakes in the western U.S. Bull. Seis. Soc. Am., 63, 501.

27. Sykes, L.R., and M.L. Sbar (1973), Intraplate earthquakes lithospheric stresses and the driving mechanism of plate tectonics. Nature, 245, 298.

28. Teng, T.L. and T.L. Henyey (1974), Microearthquake monitoring in the City of Long Beach area for the year 1973. U.S.C. Geophysical Laboratory Technical Report #74-1.

29. Teng, T.L., C.R. Real, and T.L. Henyey (1973), Microearthquakes and water flooding in Los Angeles. Bull. Seismic Soc. Am., 63, 859.

30. Thatcher, W. and J.N. Brune (1971), Seismic study of an oceanic ridge earthquake swarm in the Gulf of California. Ray Astro. Soc. Geophys. Jour., 22, p. 473-489.

31. Trifunac, M.D. and A.G. Brady, (1976), Correlation of peak acceleration, velocity and displacement with earthquake magnitude, distance and site conditions, Eartquake Eng. and Structural Dyn., 4, p. 455.

32. Wallace, R.E. (1970), Earthquake recurrence intervals on the San Andreas fault, Geol. Soc. Am. Bull. 81, p. 2875.

33. WASH 1254, J.A. Blume, R. Sharpe, and J. Dalal (1973), Recommendations for shape of earthquake response spectra. Directorate of Licensing, USAEC Contract #AT(49-5)-3011.

34. Ziony, J.I., C.M. Wentworth, and J.M. Buchanan (1973), Recency of faulting. A widely applicable criterion for assessing the activity of faults. World Conf. on Earthquake Engineering, Fifth (June, 1973), Rome, Italy, p. 1680.

3

ESTIMATED UNDISCOVERED RECOVERABLE
OIL AND GAS RESOURCES OF THE CONTINENTAL SHELF
OF THE UNITED STATES

by

E. W. Scott
U.S. Geological Surveys

Assessment of undiscovered oil and gas resources is an important function of both industry and government and it is vital to the planning process in both areas. Many resource estimates have been made in the past, but owing to the use of various methodologies and diverse data banks there have been wide differences in these estimates. An example of the problem is shown by a table (Fig. 1) from a 1974 Oil and Gas Journal article (West, 1974), that gives four estimates of undiscovered recoverable oil and gas of the onshore lower 48 states that range from 9 to 214 billion barrels of oil and 65 to 1,177 trillion cubic feet of gas.

In September 1974, the Resource Appraisal Group (RAG) of the Branch of Oil and Gas Resources, U.S.Geological Survey was asked to aid the Federal Energy Administration in its legal responsibility to generate an independent appraisal of the undiscovered oil and gas resources of the United States, both onshore and offshore. It was stipulated in the charge to the Resource Appraisal Group that the primary emphasis was to be placed on conventional oil and gas for all onshore provinces and for offshore provinces to a water depth of 200 meters. Excluded from consideration were oil shales, tar sands, and heavy hydrocarbons and tight gas sands that were not productive at the time.

Opposing views of Lower 48 onshore potential

	Undiscovered recovered oil (billion bbl)	Undiscovered recoverable natural gas (trillion cu ft)
U.S. Geological Survey	110–214	593–1,177
National Petroleum Council	53–701	550
Mobil Oil Corp.	11	65
M. King Hubbert	9	100

Fig. 1   (after West, 1974)

32

The results of the project, which were reported in U.S. Geological Survey Circular 725 (Miller et al., 1975), are presented here, and to give them more meaning I present background information about the project to show how the figures were derived and to indicate some of the weaknesses of such estimates. I will also describe some further plans of the Resource Appraisal Group.

It is essential that the meaning of "resource" is understood. Figure 2 is a box diagram showing the relation between Resource and Reserves as used in Circular 725. Resources are defined as commodities in such form that economic extraction is currently or potentially feasible. Reserves are those quantities of resources that have been identified and can be extracted under current conditions of economy and technology. The stippled area in the upper right part of the box represents the resources that were estimated by the Resource Appraisal Group and that are the subject of this paper - Undiscovered Recoverable Resources. As indicated in this figure, it is assumed that the deposits of oil and gas would be economical if found.

Early in the resource appraisal program, the United States was divided into 15 geographical regions (Fig. 3), 11 onshore and 4 offshore. The majority of the regions consist of two or more individual geologic provinces, which are the basic elements for the appraisal. A total of 102 separate provinces were individually appraised. The areas of specific interest in this paper are the four offshore regions indicated by the suffix A on Fig. 3. The four regions include 27 separate geologic provinces as follows:

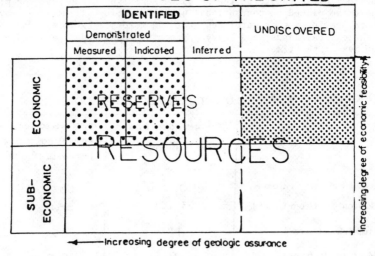

PETROLEUM RESOURCES OF THE UNITED STATES

Fig. 2. Diagrammatic representation of petroleum reousrce classification by the U.S. Geological Survey and U.S. Bureau of Mines (modified from McKelvey, 1973).

Fig. 3. Index map of North America showing the boundaries of the 15 regions
and offshore, covered in U.S.G.S. Circular, 725 (after Miller et al.,
1975)

Region	1-A	Alaska Offshore	13	provinces.
Region	2-A	Pacific Coast Offshore	9	provinces.
Region	6-A	Gulf of Mexico	1	province.
Region	11-A	Atlantic Coast Offshore	4	provinces.

The Federal Energy Administration had a legal responsibility to have the appraisal
ready by June 1975 and as a result of this urgency, the Resource Appraisal Group
obtained the assistance of more than 70 geologists within the U.S. Geological Survey who had broad experience or expertise in particular areas or provinces that
were to be evaluated.

Geologic data formats were devised and distributed to the geologists making the
preliminary geologic appraisal of a specific province. The format consists of 85
basic categories of information that would allow the geologists to provide data
needed to characterize the basic geology of the province. In short, the format

provides the basic input essential to the various methods of appraisal that would
be applied by the Resource Appraisal Group.

The completed geologic data sheets, and production and reserve data where available, were subjected to geologic and volumetric-yield analog procedures to determine a range of hydrocarbon yield values. Other procedures were also used, such
as extrapolating known producibility into untested portions of a province.

Finally, all published and documented resource appraisals were compiled and appraisals were made for each province by a comprehensive comparison of all of the information. Then, assuming that the oil and gas would occur in commercial quantities, an initial resource appraisal was made by a subjective probability technique
as follows:

> A low resource estimate corresponding to a 95% probability that
> there was at least that amount.
> A high resource estimate with a 5% probability that there was
> at least that amount.
> A modal estimate which the estimater associates with a most likely
> amount.
> A statistical mean which is derived by adding the high, low, and
> modal and dividing by the three.

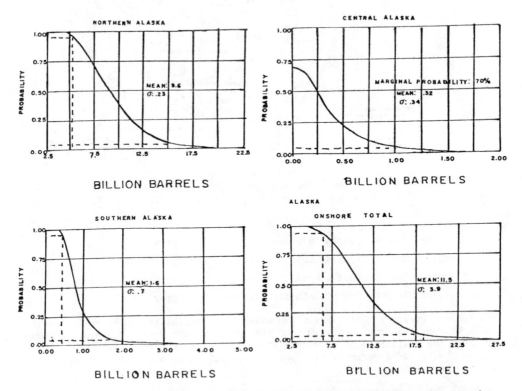

Fig. 4. Probability distributions by Monte Carlo analysis on undiscovered
recoverable resources for Alaska: Aggregate probability distributions for three onshore subregions and the total Alaska onshore.
(after Miller et al., 1975).

Long normal probability distributions were computed for each province in order to translate the separate probability judgments into a useful form for evaluating resource assessments for total regions and the whole United States. Once the assessments for each province were fitted with a distribution, the probability function in each of the 15 regions was computed by Monte Carlo techniques. This was followed by aggregations for onshore and offshore areas and the whole United States, giving the ranges and mean values of estimated resources.

Graphic displays were made for the probability distributions of the undiscovered recoverable resources by province (or subregion) and region and totals for the United States. An example is shown in Figure 4 where curves are displayed for oil estimates for the three onshore subregions of Alaska and the total for the Alaska onshore.

Estimations of undiscovered oil for the offshore by regions, in billions of barrels, are:

	95% Probability	5% Probability	Statistical Mean
Region 1-A Alaska	3	31	15
Region 2-A Pacific	2	5	3
Region 6-A Gulf of Mexico	3	8	5
Region 11-A Atlantic	2	4	3
Total Offshore United States	10*	49*	26*

Gas estimates were derived in the same manner as described for oil. The gas estimates for the continental shelf, in trillions of cubic feet, are:

	95% Probability	5% Probability	Statistical Mean
Region 1-A Alaska	8	80	44
Region 2-A Pacific	2	6	3
Region 6-A Gulf of Mexico	18	91	50
Region 11-A Atlantic	0	22	10
Total Offshore United States	42*	181*	107

* Totals for high and low values are not obtained by arithmetical summation; they are obtained by statistical methods.

These appraisals show that nearly 32% of the undiscovered oil and 22% of the undiscovered gas of the United States lies beneath the continental shelf, thus giving some measure of the importance of the offshore area as a possible source of much of our future oil and gas supply.

It was indicated above that there were some weaknesses in such estimations. The numbers presented were derived by application of subjective reasoning to available geologic data and these estimates can go up or down as new data are acquired. Further, there is no way at present to determine the presence of commercial accumulations of hydrocarbons without drilling. A good example is the Mafla area (off-shore Mississippi, Alabama and Florida) of the Gulf of Mexico. All exploration efforts in the area indicated the presence of suitable geologic conditions that could entrap billions of barrels of oil or trillions of cubic feet of gas and the petroleum industry backed this thinking by spending nearly $1.5 billion for federal leases. Subsequently, 18 expensive tests have been drilled without finding commercial accumulations and nearly half of the tracts that were leased have now been dropped.

Similar, but not as costly, examples are in the Gulf of Alaska and the southern California borderland. A review of the latter area shows that since 1971, there have been several published estimates of undiscovered resources for the offshore area of southern California. These estimates have been widely divergent, but an accurate comparison cannot be made because of the different methods of reporting. Some are given as oil-in-place whereas others are given in terms of recoverable resources.

Most estimates have been based on figures given in the American Association of Petroleum Geologists Memoir 15 - Future Petroleum Provinces of the United States; Their Geology and Potential (1971). The paper on southern California by Parker (1971) gives an estimate of in-place oil of 75 billion barrels for 48,000 square kilometers of the borderland. The reader was cautioned, however, by Parker's concluding statement that the figure -

> "-- is a fair estimate if the thickness and extent of the
> Pliocene and upper Miocene are as shown on the maps and
> sections. However, these control factors are based on
> scanty information and liberal interpretations."

The Western Oil and Gas Association (1974) applied a "typical" recovery factor for California fields to Parker's estimates and concluded that the recoverable resources within the borderland would be in the range of 16-19 billion barrels with 14 billion barrels as a reasonable estimate. Also in 1974, the U.S. Geological Survey made an estimate for the Bureau of Land Management for a sale area covering 6200 square kilometers or about 13% of the total borderland. This estimate had a range of 1.6 to 2.8 billion barrels of recoverable oil (Bureau of Land Management, 1975, p.1). Taylor (1976) cited the figures presented in 1975 by the Resource Appraisal Group and pointed out that there were not sufficient new data to alter these estimates.

Why had there been such a large reduction in the estimated amounts of hydrocarbon resources from 1971 to 1976? Part of the answer was indicated in Parker's quoted statement regarding the uncertainties. After 1971, many kilometers of seismic lines have been run, many bottom samples and shallow cores have been taken, and a 3,000-m stratigraphic test well has been drilled. All of this has added a great deal to the geologic knowledge of the southern California borderland. The Pliocene - upper Miocene section is important in relation to the presence and amounts of available hydrocarbons, because 87% of the hydrocarbons produced to date from the five basins have been from this restricted part of the geologic column (Taylor, 1976).

A quick review of the geology of southern California shows a Pliocene - upper Miocene section in the Los Angeles Basin more than 6,000 m thick. Moving seaward, southwest from the Los Angeles Basin, this important objective section thins rapidly and is absent on the crest of the Tanner Banks - Cortes Banks area of the southern California borderland. It was there that federal leases were obtained by various oil companies, and five tests have been drilled without any reports of significant shows of oil or gas.

This summary history of the southern California area shows the importance of adequate and up-to-date geologic information in any program of resource estimation. Great amounts of data have been acquired in recent years over much of the offshore area of the United States, and future plans for the Resource Appraisal Group include a complete review and update of these areas through acquisition and analysis of available data and by development and integration of new methods. The program will also be expanded into deeper water and into new areas.

The proposed program of updating and expansion places a high priority on the frontier areas of the continental margin in light of the fact that a large part of the oil and gas needed to meet our future energy demands will come from some of these areas.

REFERENCES

1. Bureau of Land Management, 1975, proposed 1975 outer continental shelf, oil and gas general lease sale, offshore southern California, DCS Sale No. 35: Draft Environmental Statement DES 75-8, 4 vols., variously paged.

2. Miller, B. M., Thomaen, H. L., dolton, G. L., Coury, A. B., Hendricks, T. A., Lennartz, F. E., Powers, R. B., Sable, E. G., and Varnes. K, L., 1975, Geological estimate of undiscovered recoverable oil and gas resources in the United States: U.S. Geological Survey Circular 725, 78P.

3. Parker, F. S., 1971, Petroleum potential of southern California offshore, in Cram, I. H., ed., Future petroleum provinces of the United States - their geology and potential: American Association of Petroleum Geologists Memoir 15, v.1, p. 178 -191.

4. Taylor, J. C., 1976 Geological Appraisal of the petroleum potential offshore southern California: U.S. Geological Survey Circular 730, 36 P.

5. West J., 1974, U.S. Oil -Policy riddle: How much left to find? Oil and Gas Journal, Sept. 16, 1974.

6. Western Oil and Gas Association, 1974, Environmental assessment study, proposed sale of Federal oil and gas leases, southern California outer continental shelf: Los Angeles, Dames and Moore, Inc., 3 vol.

4

OFFSHORE FRONTIER BASINS OF THE PACIFIC STATES AND ALASKA:
SOME GENERALIZATIONS ABOUT THEIR GEOLOGIC SETTING AND
PETROLEUM POTENTIAL

by

David W. Scholl
U.S. Geological Survey

Thick accumulations of sedimentary rocks, mostly of Mesozoic and Cenozoic age, underlie major portions of the continental margin (i.e. shelf and slope) and adjacent deeper-water areas that border California, Oregon, Washington, and Alaska. Some of these sedimentary basins have been partly explored, but most of them are undrilled or frontier basins. Many of these basins probably formed in response to a variety of tectonic circumstances related to the interactions (both direct and indirect) of major lithospheric plates. Regional geologic considerations and the implications of publicly available geologic data allow for speculations about the likelihood that offshore Pacific-Arctic basins contain significant hydrocarbon deposits.

Onshore coastal basins in southern California contain large amounts of oil and gas in thick turbidite beds of chiefly Miocene and younger age; smaller amount occur in other reservoir rocks. The turbidites accumulated in extensional basins that formed during the past 10 - 15 m.y. in response to strike-slip or transform motion between the Pacific and North America plates. Offshore, however, equivalent rock sequences in similar basins are much thinner in all but the more inshore (landward of the Santa Rosa-Cortez Ridge) basins of the southern California Borderland, and possibly the Eel River Basin off northern California. Exploratory drilling in some of these basins, and also in underlying Tertiary and Mesozoic beds that form part of their structural framework, has discouraged speculations that large volumes of hydorcarbons occur in frontier areas offshore of California.

The continental shelf and slope that borders Oregon and Washington is underlain by more than 5 km of Eocene and younger sedimentary and volcanic rocks. This sequence has been deformed in response to episodic periods of underthrusting and transform faulting between Pacific and continental lithosphere. The results of sparse exploratory drilling, which failed to find adequate reservoir beds, suggest that the thick sub-shelf sections do not contain large deposits of hydrocarbons. However, maverick ideas about the likely location of potential reservoir rocks, based upon studies of the distribution of deltaic facies, and of the reservoir potential of large infolded masses of sandstone in melange sequences, have not been tested.

Like the Washington-Oregon margin, the continental margin of the central and eastern Gulf of Alaska is underlain by a great thickness (as much as 10 km) of Cenozoic sedimentary beds that have been compressionally deformed by the underthrusting of the Pacific plate. Exceptionally high rates of sedimentation have been maintained here during the past 5- 10 m.y. because of mountainous coastal relief and alpine glaciation. Although large folds are present in the sub-shelf sequences, regional mapping and discouraging results of offshore drilling suggest that suitable reservoir beds in the thick Miocene and younger deposits are not structurally or strat-

igraphically proximal to middle Tertiary rocks that are widely believed to be potential source beds. Perhaps lower Tertiary rocks, which formed under somewhat different tectonic conditions, may contain adequate reservoir units, a possibility
that would enhance the oil and gas potential of the eastern and central Gulf of
Alaska. However, evidence of secondary mineralization and the results of drilling
near Middleton Island weaken this speculation.

The five or six frontier basins of the western Gulf of Alaska (Middleton Island to
Unimak Pass) are underlain by a 2 - 6 km thick layer of Neogene beds overlying
older and more deformed Tertiary and Mesozoic sequences. Neogene beds near Kodiak
Island may be too thin to have generated hydrocarbons, but underlying lower Tertiary
beds may have supplied oil or gas to Neogene reservoirs. Farther west, in the
vicinity of the Shymagin and Sanak Islands extensional rifting has formed grabens
that contain thicker Neogene sequences. Little is known about their petroleum
potential.

At least five very large frontier basins occur west of Alaska on the Beringian
shelf, a broad epicontinental sea that covers at least five major basins. These
basins (Bristol, St. George, Navarin, Anadyr, and Norton) contain mostly Cenozoic
deposits that are as much as 15 - 20 km thick, although Cretaceous beds are likely
in some of them. Navarin and St. George along the southern edge of the shelf,
probably formed in response to crustal extension that began 60 -70 m.y. ago when
formation of the Aleutian arc terminated oblique subduction of Pacific crust beneath the Beringian shelf. Although factual information is sparse, regional
considerations suggest that the Beringian basins may be excellent prospects for
significant oil and gas resources.

Another prospective area, the deep-water (3 - 4 km) Bering Sea Basin, also lies
off western Alaska. This marginal oceanic basin is underlain by oceanic crust,
evidently a sector of Pacific crust that was trapped here when the Aleutian arc
formed in latest Cretaceous or earliest Tertiary time. These crustal rocks are
buried beneath a 4- to 10-km-thick blanket of sedimentary deposits that is mostly
of Cenozoic age. Evidence from geophysical data and offshore drilling suggests
that hydrocarbon source and reservoir beds may occur although the hydrocarbons may
prove difficult to recover for technological and economic reasons.

The frontier basins mentioned thus far formed on relatively young crustal rocks of
Mesozoic and Cenozoic age. Unlike them, the frontier basins of the Chukchi and
Beaufort Seas (e.g. Hope, North Chukchi and Camden Basins) off northwestern and
northern Alaska, respectively, formed on ancient continental crust that includes
basement rocks of Precambrian and early Paleozoic age. These rocks are part of
the Franklinian sequence of the Canadian Arctic, an area that apparently supplied
sedimentary debris to northern Alaska until seafloor spreading in the Late Jurassic
formed the Chukchi and Beaufort continental margins by rifting northern Alaska from
northwestern Canada. Substantial thicknessess (3 - 10 km) of moderately or only
slightly deformed late Mesozoic and Cenozoic sedimentary deposits have since
accumulated beneath these subsiding shelf basins. Significantly, roughly equivalent rock sequences are known to contain promising oil and gas deposits in northern Alaska and the MacKenzie Delta area. Moreover, the unusual structural-strat-
igraphic relation that makes the Permian and/or Triassic section at Prudhoe Bay a
giant oil and gas accumulation may also extend seaward beneath the inner part of
the western Beaufort shelf and the northern Chukchi shelf. Accordingly, optimism
is justified that large offshore petroleum accumulations may occur in Tertiary,
Mesozoic, and upper Paleozoic rocks underlying the continental margins of the
Chukchi and Beaufort Seas.

SECTION II

<u>NOVEL ENERGY AND RESOURCE RECOVERY METHODS</u>

Energy and the Oceans : Myths and Realities

Novel Methods for Recovery of Marine Kerogen and Bitumen

Urban and Fish Processing Wastes in the Marine Environment:
a Case of Wasted Energy at Terminal Island, California

A Comparison of the Alcan Plan with the Methanol Approach
for Prudhoe Bay Natural Gas

5

ENERGY AND THE OCEANS: MYTHS AND REALITIES

by

Bernard LeMehaute
Tetra Tech, Inc.

A storm of gigantic proportions is pounding our United States coastlines. This storm is not the result of an extreme meteorological event, but it will be with us for a long time. It is the result of the political and legal fights to determine which areas should be converted into industrial complexes, recreational resorts, or kept in their pristine state.

The fact that the shoreline is under stress is an understatement. It is forecasted that more than 60% of Americans will live near the coastal water by 1985.

Nearby, at Point Conception, is a typical example of the dilemma. One can understand why the people who live on Hollister Ranch want to prevent the construction of an LNG terminal at Cojo. On the other hand, we have seven million people in Los Angeles who need gas, energy and jobs.

And the pressure remains, from the population to live closer to the beach, from the industries to have access to the sea for myriads of technical and economic reasons.

The solutions are complex and will not satisfy every group.

It is the function of the engineer to bring to light technological and scientific facts and numbers, as much as possible, in such a way that the economists, the politicians, and the population at large are allowed to make a choice "en connaissance de cause". I will leave to my colleagues tomorrow the most difficult task to treat the socio-legal implications.

From an engineering point of view, the problems facing coastal development and preservation are multiple. The loss of real estate value by shoreline erosion amount to billions of dollars each year. The delay in construction of superport-Loop in Louisiana is finally being done--and the legislative process which is needed to establish any kind of job-producing industries along our shorelines are dragging our productivity and the economy.

I would have liked to present the engineer's point of view to all these problems and propose possible solutions and compromises which may accommodate both the economy and the ecology.

These solutions are not straightforward. Because of the small amount of time that we have together, we will limit our discussion to a few topics. Considering its importance, and the present interest that most people have in the subject matter, we will talk about water and energy. This may not be original, and so much has already been said, but I hope that I will bring to you some facts which I believe may have been overlooked.

---

*The material of this paper has appeared under the title "Extracting Energy from the Ocean" in Managing Ocean Resources: A Primer, edited by R. L. Friedheim, Westview Press, in pp. 51-66.

During this brief exposure on what is a monumental subject, I would like to survey the multiple facets of the interrelationship between ocean water and energy. We will examine both water and in particular, ocean water as a source of power and water usage in energy production. We will also survey the subsequent problems which are created and the ultimate solutions.

In the past, dependence of food and energy production on the availability of water was localized; but today, this dependence is so strong that an integrated approach is absolutely necessary.

The water needs vary from region to region. Globally, if one takes into account variations in specific consumption due to domestic, agricultural and industrial needs, one estimates that there is enough land and fresh water on earth to sustain a population of 30 billion people. The North American continent can provide a comfortable 1020 Km^3 of water per year per capita to 1.145 million people.

The degree of interdependence between water, food and energy varies over the world from countries such as Canada, with its small density of population and vast water resources, to countries of the Middle East, where water is a most precious commodity. In many states, the water consumption and future demand already exceed the local supply, and the future energy demand reinforces the problem.

In general, the country is water-rich east of the Mississippi River and water-poor to its west with the exception of the water-abundant Pacific Northwest. Even within the water-rich regions though, there are localized supply problems. The Boston-Washington megalopolis and parts of Florida are examples. From an overall standpoint, however, the East is not as seriously constrained by its lack of water supply as by the environmental problems besetting it. "Water, taken in moderation, never hurt anybody," said Mark Twain. Today's reality is, "Water used in immoderation hurts everybody."

Problems associated with water usage are critical worldwide. These problems, however, are solvable. The solutions to these problems do not require new breakthroughs in technology, but rather new approaches in technical management and a clear understanding between the fundamental relationship between the ocean's water and energy requirements.

Water is essential in all human and industrial functions. It takes, for example, 20,000 gallons of water to produce the steel for one car and 10,000 gallons for the car's assembly. Also, much more water is needed for food and energy production.

For thousands of years, man has recognized, sometimes only dimly and sometimes clearly, that water moves in a continuous self-purifying cycle. This is the cycle that begins when rain falls from clouds, sinks into the earth, seeps into river, lakes and oceans, then evaporates into the atmosphere to begin the cycle all over again. In one vitally important aspect, water's behavior is steadfast; the world's supply of water neither grows nor diminishes. It is believed to be the same now as it was 3 billion years ago.

As Don showed us this morning, the oceans, ice caps, and glaciers contribute 99.35% of the earth's water. The small remainder accounts for all the earth's rivers, lakes and underground water tables. Yet it is this small amount that we use every day in myriad ways.

About 517 trillion metric tons (equivalent to a volume of 124,000 cubic miles of water) evaporates into the atmosphere annually. A far greater part, about 88%, rises from the ocean. 12% is drawn from the land, evaporated from lakes, rivers, streams, etc. Of the water that goes into the atmosphere, 79% falls back directly into the oceans. Another 21% falls on the land, but 9% runs off into the rivers and streams and is returned to the oceans within days. The remaining 12% soaks into the land and is availabel for plant and animal life processes.

In these processes, too, water intake matches output, as animal and vegetable life exhales, excretes, and perspires what was earlier ingested through root and mouth. In a sense, the world's water circulation system operates like a complex, gigantic pump driven by solar energy. Endlessly, recycled water is used, disposed of, purified and used again by nature including humans. The coffee that you drink at noon may have been made of what was, ages ago, the bath water of Archimedes. Past, current and projected growth of the world's population with its huge demands for energy and food has and will profoundly affect this natural cycle in terms of biological, chemical and thermal waste (while chemical waste can be processed, thermal waste cannot).

The increasing demand for water supply will come from the increasing demand for energy production and food supply. Both are dependent upon energy availability.

The first form of energy extracted from the water element is hydro power. Hydro power has not been given the proper attention in view of our energy shortage. It is a nonpolluting form of energy. There are still huge amounts of power readily available which could be obtained by harnessing rivers and tidal embayments even in the United States.

The U. S. has the potential to quadruple its production of energy from hydro power - still excluding energy rich Alaska. The U. S. Army Corps of Engineers has identified 49,500 existing dams which could potentially be considered for hydropower production. 3,000 of these dams are in energy poor New Hampshire. Some of them were used in the past for hydropower, but their operation was stopped as being non-competitive in an oil-cheap era. The harnessing of many of these installations would be economical and should not create any environmental impact since the dams are already built.

The potential for new hydropower installations is also significant. The economic problem sometimes arises from transporting the energy from the center of production to the center of consumption. A few worldwide examples are:

- The Inga Project considered in 1950 consisted of harnessing the natural fall of the Congo River with 300 turbines of 200 kw each. This energy would then be transported to Europe through 400,000 v. D.C. lines.

- The torrents of the Himalay Mountains are practically in their pristine state while mere penstokes and Pelton turbines can provide India with all its needed energy. But the mountains are in the North and the centers of consumption are in the South. Nevertheless, is there any need for India to resort to nuclear power and discard hydropower so easily?

- Tidal power availability at San Jose, Argentina is practically illimited. The center of consumption is Buenos Aires, further North.

- In the United States, a tidal power project at Cook Inlet can provide us with 75,000 million Kwh per year, while the demand in Alaska is only 1,200. It is about 3% of the total U. S. Electrical production and 1% of the total U. S. Energy consumption.

I would not like to speculate, at this time, on a possible breakthrough in energy transportation (super conductors, transmission of hydrogen, etc.), but if it were to happen, these projects and many others of this type would become immediately viable.

Since we have talked about tidal power, let us now survey the ocean as a potential source of energy because one heard so many misleading statements about the subject matter.

It is true that the amount of energy potential available in the sea, both visible and invisible, is practically limitless and thermally non-polluting. Some of the estimates of this energy are truly staggering:

25 billion kw of energy is constantly being dissipated along the world's shorelines by waves. A surface wave three meters high with constant period and amplitude transmits 100 kw for every meter of wave crest of 100 megawatts per kilometer (assuming that the wave period is 11.3 seconds). It could be compared to the power of a line of automobiles, side by side, at full throttle. Tides could be harnessed to produce 1240 billion kwh per year.

The Gulf stream off Florida has a volume flow of over 50 times the total discharge of all the rivers of the world.

Enough electricity could be generated from heat engines operated on the temperature differential between surface and deeper ocean waters to provide more than 10,000 times the world's yearly electric power requirements.

The salinity power represented by the osmotic pressure between freshwater and salt-water which is equivalent to a water fall of 700 feet. From the global runoff of freshwater into the oceans, 2.6 billion kw of power can potentially be obtained.

Energy from the ocean has many forms. There is enough solar energy to maintain the entire world population at an affluent level, but it is a diffused form providing at best 100 watts/square foot in full sunlight. Some means for collecting this energy must therefore be found.

Wind power is one form which has been the major source of energy used for ocean transportation until the invention of the heat engine. Wind mills are also extensively used at special sites, but a steady velocity of 30 to 40 knots is required for the economic generation of energy. The worst place to install a wind power generator is at sea despite stronger prevailing winds. On land, about one third of the cost is the tower, two thirds the propeller and the auxiliaries. At sea, the supporting structures must not only stand the wind force but also the wave force, so increasing its cost by an order of magnitude.

In order for an energy production system to be worth developing, it must satisfy a number of criteria. Three of them are:

1. it must use a high energy density medium
2. it must have a simple processing method
3. it must use a source of energy which is dependable.

The energy density is directly related to the volume to be processed and to the size of the machinery which is needed to extract the energy from that medium. Therefore, the cost is related to energy density.

For example, oil is a high energy density medium. When one deals with hydropower, the energy density is proportional to the head: the higher the head, the cheaper the energy. One of the cheapest forms of energy is the one provided by the water power coming from mountains through narrow penstocks and ejectors and splashing the buckets of a simple Pelton turbine. As the head decreases, one has to switch to the more bulky Francis turbines. The cost of energy generally increases as the head decreases, and there is a limit which is considered economically unattractive. When the head is below, say 2 meters, even a small turbine built to satisfy local needs could not be considered economical.

By the same token, one can disregard a priori all the energy which is contained in large currents such as the Gulf stream. Typically, the speed of the Gulf stream is 0.6 to 2 m/sec, the corresponding maximum head is 20 cm. It can, in no case, enter the realm of economic expectations.

The total amount of power available in water wave is also huge, but it is very diffused. The energy density (i.e., per unit of volume at the free surface) or equivalent head is so small that it can only be considered for local and special purposes as buoys and in isolated areas.

The place of the tidal power in terms of energy density is easy to assess: the head is linearily related to the tidal amplitude and the energy available is proportional to its square. Tidal heads are generally at the lower limit of what should be considered economical. Therefore, tidal power in general is a marginal form of energy to be considered at special sites with narrow bay entrances (to limit the cost of dike construction).

The million kw Passamaquoddy project in the Bay of Fundy would cost 3 billion dollars nowadays because of the length of the dikes. But a small, entirely U. S., project at Cobscook Bay seems economical because of its narrow entrance.

The processing methodology has a significant impact on cost. The simpler it is, the least capital cost and the least maintenance cost. The transformation of water power energy into mechanical and electrical energy is relatively simple. It is a very effective process. The efficiency of a hydraulic turbine is as high as 96%. Water power is one of the simplest energy process conversions which exist, by opposition to thermal, chemical or nuclear conversion.

The energy density over a thermal gradient is very large. Each 1 $^{\circ}$C of temperature difference corresponds to an equivalent head of 426 meters, implying a potentially high energy density. The problem is processing.

The ocean thermal energy conversion is presently considered with great optimism by many. The earth's oceans can be viewed as natural thermal collectors and storage devices. At 1500 feet below sea level, the water temperature is close to the freezing point and on the surface, the water temperature in tropical regions range between 70-85 $^{\circ}$F. These temperature differences cause density stratification that retards fluid mixing and prevents the ocean from arriving at a uniform temperature. The existence of the temperature gradient in principle can be tapped to run heat engines, to generate electricity or hydrogen.

The thermodynamic efficiency of ocean thermal energy conversion (OTEC) power plants is controlled by the available ocean water temperature differential, efficiency of the heat exchangers, fluid energy losses from pumping large amounts of water,and frictional losses incurred by the Carnot efficiency. This astonishingly low efficiency does not preclude its viability, however, since the fuel, in this case seawater, is free. But the processing methodology requires a huge amount of liquid to be circulated through a heat exchanger. For this reason, there is very little hope that this process may become economical in the near future, but its cost may become competitive if the cost of fossil fuels continue to escalate. Nevertheless, it will take many years of OTEC platform. However, significant improvements in the heat transfer surface technology could make the OTEC concept appear in a more favorable light. The cost of 30 mills per Kwh has been presented as feasible.

The benefit of the present OTEC project is multiple. The technology fallout in ocean engineering as the result of the research presently being done could be significant. For example, there is some hope that interesting solutions can be found for biofouling control, deepsea mooring, construction of large structures, and special shipyard development. The DOD may also benefit from such a project since the OTEC platform goes deep into the sound channel. But there is very little hope in the near future that the banks will one day loan money to a utility company or the stockholders invest capital to build an OTEC platform. The total power generated by OTEC, like tidal power, will remain very small in terms of world energy demand. Our effort should be pursued, but our expectation realistic: at most 1% of our energy supply by the year 2000.

The same consideration applies for energy from osmotic pressure generated by the mixing of fresh water with seawater. This amount of energy is also huge. The head is 700 feet. The scheme to recover this energy is relatively simple since it is sufficient to build an underground water power plant at the mouth of rivers and to connect the tailgate tunnel with a network of underwater osmotic membranes spreading the freshwater in the ocean. The problem is then that the area of membranes per unit of fluid discharge is so large that it does not appear economically feasible. However, basic research in this field should be pursued at a moderate rate of expense.

Therefore, from all the sources of energy available in the oceans, tidal power is the one which necessitates the simplest processing method. It is to be expected that the wear and fatigue of other sources of energy at sea such as wave power or thermal gradient will be such that their lifetime may be shorter since they are exposed to the violence of the oceanic elements. By opposition, tidal power installations are to be built in well-protected areas. Tidal power once installed could still be there centuries from now. Some old tidal mills still exist in Europe and they were built centuries ago. France, USSR and China have tidal power plants and others are being planned in Korea and Canada.

Finally, the source of energy must be dependable. Oceanic thermal gradient, and osmotic pressure energy from rivers arriving into the oceans are dependable sources of energy. The recovery of energy from wave power has always fascinated inventors, but it is not dependable. R. Dhaille reported in 1956 to have examined more than 600 patents, some of them demonstrating a disarming ignorance of the laws of physics. He concluded that not even one was worth developing. More than 80 years ago, Albert Stahl cited 20 proposals for harnessing wave power in the United States. The back and forth motion of breaking waves and at times their devastating effects may give the impression that a huge amount of energy could be tapped.

The main economic problem arises because any structure has to be designed to withstand the largest storm waves, while the prevailing waves, which would provide most of the energy, are smaller by an order of magnitude. Finally, wind-generated waves are not a dependable source of energy as a result of the large variation of sea states inherent to meteorological fluctuations.

Along the United States Pacific Coast, the 100 kw/m is exceeded 1% of the time at most locations, and 2.5 kw/m is exceeded 80% of the time only.

The most ambitious project now being considered for recovering wave power is located 10 miles west of the Hebrides Islands off the coast of Scotland. The proposed device would be a series of axles each with 20 to 40 huge swiveling cams. Each set would be as long as a supertanker. The set of cams will then be a few hundred kilometers long.

The up and down wave-induced motion pumps water to high level. The corresponding
potential energy is then processes through a turbine-generator or is used to generate hydrogen. Each cam unit would cost $48 million and will produce 50 mega-
watts based on the severe nature of the sea states in this part of the world. Its
promoter considers wave power as the ultimate clean solution to the electrical
requirements of the United Kingdom. Whether this optimistic assessment is substantiated by the facts remains open.

At the opposite, tidal power is extremely dependable since the tidal amplitude can
be predicted as far ahead as needed as a function of trajectories of celestial
bodies. Still tidal power fluctuates substantially from the equinox to the solstice
and from day to day. More importantly, the sea level varies continuously with time.
But there are ways to cope with that variation, in such a way that one can produce
the energy at the time that it is required. In this respect, tidal power is very
dependable. It has been said that tidal power varies with the moon when the life
on earth is regulated according to the sun. Actually by various schemes, it is
possible to extract tidal power at any time one wants, either a peak power or
steady production. In this respect, the tidal power can be classified as one of
the most dependable forms of energy which is available even more dependable than
water power from rivers since floods and dry periods are to some extent also un-
predictable.

Furthermore, there is very little maintenance to be done once it is built. The
relatively large capital cost which is required initially has to be balanced by
the fact that after it is built, it will be there forever. By opposition, a thermal or nuclear power plant has a 30 year lifetime. But the United States has very
few sites which could be harnessed economically.

As one can see, there is very little hope that "Project Independence" could be
achieved from energy sources from the water element alone.

Let us now examine our need for water and to begin with, the demand that other
energy sources create for water.

In the United States, most energy processes are dependent on uses of water. (By
contrast, in the Middle Eastern countries, most water problems are dependent upon
uses of energy). Oil, natural gas, coal, oil shale, synthetic fuels, geothermal,
and nuclear power are all energy sources which need water.

Oil and natural gas both require water for drilling. Wherever secondary or tertiary recovery techniques are used, however, much larger amounts of water are
necessary, since vast amounts must be pumped underground to force the underlying
oil or gas to the surface. Since the United States consumes approximately 17
gallons of water for each barrel of oil produced, water is intimately related to
these forms of energy production.

Coal mining consumes water for several purposes. Access roads and coal surfaces
being worked must be watered down for dust control in both surface and underground
mines. Water is also used to wash the mined coal, transport the coal away from
the mine in slurry form through pipelines, wash up by mine personnel and to help
in revegetation of stripmined areas. The amount of water consumed varies from
4 to 18 gallons per ton of coal mined, depending on the region. Water withdrawals
also vary among regions from as little as 4 gallons to as much as 430 gallons per
ton. The water requirement of coal conversion plants are not large even in the
semiarid West, compared with the quantities devoted to irrigation. (A coal refinery processing 75,000 tons of coal per day would need 15 million gallons of
water per day. This comes out to about 15,000 acre-feet per year). But a supply
of water, like a supply of coal, must clearly be taken into consideration for each

plant.

Extracting crude oil from shale with presently available technology requires large amounts of water. First of all, water is necessary in the mining and processing of the shale. Second, the oil present after retorting (distilling by heat and pressure in the presence of a catalyst) is of such high viscosity that a hydro-cracking process is necessary to upgrade the shale oil into a useful fuel. Finally, more water is needed to compact and stabilize the spent shale and to control dust. Besides these industrial uses, additional water for personal household use and re-lated industrial uses will be required if any significant development of oil shale resources is undertaken--since large numbers of workers would have to be imported into the now sparsely populated areas of potential oil shale production. Research indicates that a plant producing 100,000 barrels of oil a day will require 16,800 acre-feet of water each year.

Water quality requirements for shale oil vary. Although higher quality water is necessary for retorting and upgrading the shale oil and for household use by the industrial workers to be imported into the area, spent shale disposal can be ac-complished with brackish or other lower quality waters. This might be obtained from dewatering the shale mines.

Producing synthetic fuels from coal consumes water in three ways. First, water is used in the chemical processes that convert the coal into gas or liquid fuel. Sec-ond, water is evaporated in the cooling activities associated with these processes. Third, water leaves the processing site as moisture content in the coal ash and waste discharges. Consumptive use ranges from less than 10 gallons per million BTUs for those processes that produce pipeline gas. Research indicates a plant capable of providing 100,000 barrels of synthetic oil daily consumes over 5,900 acre feet of water a year.

Geothermal water requirements for a plant using the hot water technology would be very small, probably no more than 470 acre-feet a year for a 200 megawatt (MWe) plant. There are no foreseeable water requirements for the dry steam design, since the hot dry rock technology may or may not use water as its fluid.

Nuclear power generation withdrawal and consumption requirements are derived essent-ially for cooling needs. Nuclear plant water consumption factors vary with the type of plant and cooling processes used. In general, a water consumption factor of 0.8 gallons per kilowatt hour is needed. The mining and processing of the raw fuel - uranium ore - also requires the use of some water.

Our vast requirements for electrical power depend totally on the availability of cooling water from our lakes, rivers, and coastal zones. Indeed, if one excludes hydropower, which is thermally non-polluting, the purpose of all energy sources (fossil fuel, geothermal, or nuclear) is to create a high temperature, T_1. In order to obtain consumable energy, a lower temperature T_2 is also needed. The availability of cooling water is as important as the heat source. It is evident, if there is a large resource of energy at sea, it is "T_2," whilst the heat capacity of our lakes and rivers is limited. Water heat capacity is very great compared to that of other substances. It is that property that allows the California current to moderate our climate in Santa Barbara, and which also keeps the earth's surface at a relatively constant temperature.

Nevertheless, the problem of waste heat will be with us forever. The technological change in power generation cannot be expected to solve this problem for us. The United States consumes 7.3 trillion kilowatt hours a year, close to 35% of the world's energy consumption. Of this total, the vast majority comes from three fossil fuels: petroleum, natural gas, and coal. Together, these fuels represent more than 90% of the U. S. energy supply. Coal (with its important demands on freshwater for processing) and nuclear power will begin to increase their share of the energy supply from now to the end of the century, with more exotic energy forms contributing in a minor way.

Ultimately, all this energy consumed ends up as heat and increases T_2. Energy sources such as petroleum would not have contributed to the production of heat if it had not been consumed. The same will apply to nuclear energy, coal, etc. (water, tidal, wind and solar power result in no net increase in heat). Thermal pollution has been presented as the ultimate limit to energy development. Indeed the problem of thermal waste deserves the most attention because heat is the ultimate residual of societies activities, and the only one which cannot be processed. Any effort to concentrate it simply requires more energy compounding the waste heat.

On a local basis, the total waste heat released is substantial (for example about 5% of solar radiation in the City of Los Angeles). The total perturbation, on a global basis due to man's energy input, is small compared to the global heat budget (0.01% of insolation at the surface). The amount of energy received by the earth,by solar energy and the one which radiates is by orders of magnitude larger than the human made perturbation. Ultimately, the excess heat is diffused in the ocean, the atmosphere, and finally radiated to space. For this reason, the sea should be considered as a sink of almost infinite recycling capability. Thermal pollution problems are only of local nature and can be solved by the proper disposal and spreading. By contrast with the shoreline which is unidimensional, the land which is two-dimensional, the sea is four-dimensional, and it has not only the two horizontal dimensions but it also has depth, and it "stirs," adding a time-space relationship permitting heat diffusion and absorption of high concentrations of foreign elements.

As Dr. Don Walsh shows us this morning, for every square mile of land, there are two and a half square miles of ocean, five times deeper than land is high. One must consider our largest continents as islands.

Let us be for a moment, energy optimists; that is, let us assume that the energy supply is practically limitless (and the present technology - the breeder reactor - permits us to consider that it is already practically limitless). Then what would be the limit to the use of energy caused by thermal pollution? The total heat loss of the atmosphere to space, initially received from the sun is about 100,000 billion kw. Man-created heat is about 5 billion kw (i.e., 5 divided by 100,000 of the sun's contribution).

Let us assume for a while that the earth's population growth to 10 billion, all of them consuming 20 kwt/person. (The U. S. consumption is 10 kwt). Then the total man made heat will be 200 billion kw (400 times the present) it still will be only 0.2% of the earth's natural rate of heat loss. This would increase the earth temperature by less than $1/10^\circ C$. Thus, the total heat balance will hardly be affected since temperature variation of $2^\circ C$ or more have been recorded in the past. Therefore, on a global scale, the limit to the use of energy because of thermal pollution is at quite a distance even if the results of the previous calculations are erroneous by an order of magnitude. On a short term basis, thermal "pollution" may be considered as one of the great untapped resources of our oceans since fish farming is much more efficient in warmer water. Since the limit to the use of energy is quite large, the limit to the use of water is also very comfortable.

The sea is not only the ultimate sink, but it is also the ultimate source. The present need for water as required for energy production is and will be compounded by the need for agriculture, irrigation and land reclamation.

1600 million persons are starving and every year close to 100 million more have to
be fed. For this, many look at the sea as a potential reservoir of protein, and
even though much progress is done in mariculture and aquaculture, this effort still
has a minimum impact. Except for oysters and shrimp, new born aquaculture is in a
more experimental phase than in an operational one.

New trawler designs and equipment have created a revolution in fisheries increasing the catch by an order of magnitude, unfortunately leading to a depletion of
school populations. Therefore, the ultimate solution, in addition to fisheries
management, is desalinization, irrigation of arid land, and fertilizer plants
(which in turn require more energy).

It takes about 1 kwh of work to separate 100 pounds of salt from 300 gallons of
sea water. One presently estimates that this process will cost about 10 C in
terms of energy. The cost of water in Israel is 3 C/100 gallons.

Is it worth that cost?

One estimates that 700 m^2 of irrigated land are needed for each person, requiring
as an average 500 liters/day or about 150 gallons of water. These 700 m^2 of ir-
regated land provide enough crops for insuring the daily requirement of 2,500
calories per capita.

If one adds the cost of the conveyance system which may actually triple the cost,
it means that about 10 C of water/day/man will permit the entire humanity to live
in good health.

Water is cheap: 100 times cheaper than food, and a significant increase in the
cost of water is not going to have much effect on inflation or the economy, but its
availability and the availability of energy can effectively insure us of a better
world.

One can conclude that on a long term basis, given a limitless energy resource, such
as due to hydrogen fusion or breeder reactor, the T_2 of the ocean permits us to
produce an amount of energy practically limitless which in turn permits us to obtain a limitless amount of water for cultivation of our arid land.

I would not like to conclude that the world population can keep growing or even
that we can accept thermal pollution without further concern, monitoring and eventually control, but that a priori water, energy and even food are not the reasons
to limit the expansion of humanity. The problem is not technological but economical and managerial at a world scale. I thought it would be useful to throw in
this optimistic note at a time where one hears too many pessimistic statements.

The problem is a matter of management, a matter of resources management. One could
say that there cannot be (there will never be) a water shortage as such; but there
may be a shortage of water transmission systems, water-treatment plants, water-
storage reservoirs, or more generally, engineering construction. There may be a
shortage of capital, skills, ingenuity, resourcefulness, creativity and a national
will. There certainly may be a shortage of enlightened management. America is
like an ostrich with its head in the sand where water and energy matters are concerned. The water problem is basically a management is the environmental and
socio-cultural context for our future generation.

I recently read "undue faith in technology has created a completely unrealistic
sense of the limitness of the resources," meaning water. I will state instead,
" faith in technology, human inventiveness, and creativity permits us to consider

that, given time and enlightened leadership, water and energy are indeed limitless resources, and these resources are ultimately in our oceans."

6

NOVEL METHODS FOR RECOVERY OF MARINE KEROGEN AND BITUMEN

by

James I. S. Tang and Teh Fu Yen
University of Southern California

INTRODUCTION

The oceans and seas contain 2.6 x 10^{12} tons of organic matter, which is approximately equal to the world's resources of coal or peat (1). A large portion of this organic matter is entrapped among the conglomeration of mineral particles, shells and skeletons of marine organisms, and natural marine rocks forming the marine sediments. In its present state, the dissolved organic matter is inaccessible due to its prior adsorption by suspended mineral particles to form a firmly bonded, insoluble mass.

The sources of marine organic matter are atmospheric and riverine introduction of pollutants, accidental spillages, decompositional debris of marine organisms, metabolic end products from natural biota, and the numerous forms of intermediates derived from the interaction among the various decomposed products of living organism. Thus, stable terrestrial organic molecules can be treated as end products or different stages in the geochemical diagenic process of simple bioorganic molecules (2,3).

The organic matter in marine sediments such as kerogen and bitumen may yield commercial quantities of gas, oil and other fuels. If economical exploration and recovery methods can be developed, marine organic matter will play a significant role as an energy alternative in the United States.

BACKGROUND INFORMATION

Marine kerogen is the insoluble organic matter found in marine sediments. It accounts for most of the total organic content and is composed of humic acid, readily hydrolyzable carbon, and residual carbon. Marine kerogens are fossil remains which contain varying proportions of important naturally-occurring building blocks of plants and animals. During maturation, kerogen may yield commercial quantities of petroleum (4). Experimental results reveal that most petroleum is derived from kerogen (5,6).

Two general types of kerogen have been classified namely, coal- or gas-bearing kerogen (< 6% hydrogen content) and the oil shale- or oil-bearing kerogen (7% hydrogen content). Hypothetical structures of oil and gas type kerogen have been given by LaPlante (Fig. 1) (7). Gas type kerogen is almost indistinguishable from the coal type kerogen. However, it is limited primarily to gas generation since its low hydrogen content severely limits the number of high molecular weight paraffinic groups which are necessary for oil formation. Marine oil type kerogen is comparable to that found in oil shale formations and has potential for both oil and gas generation since it contains enough hydrogen to be developed into either liquid-hydrocarbons or gaseous-hydrocarbon fuels.

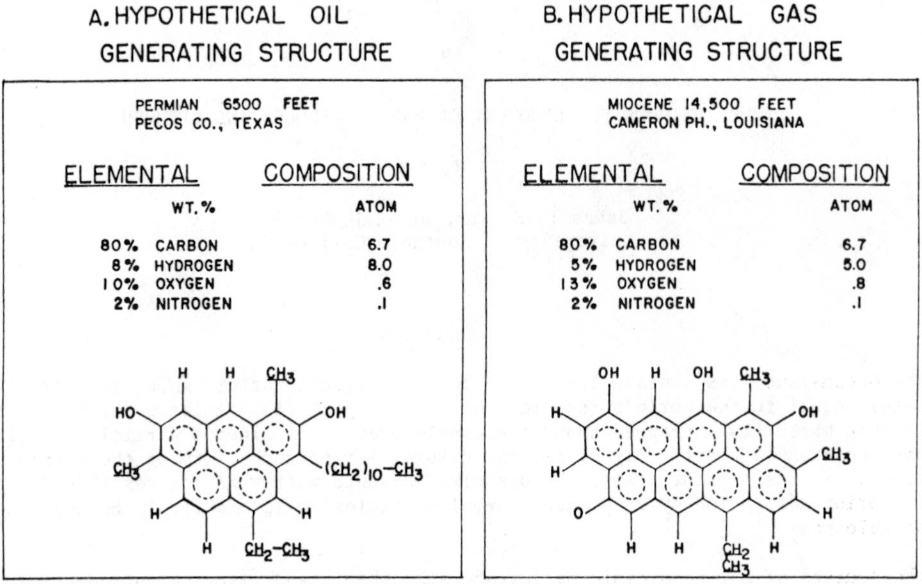
A. HYPOTHETICAL OIL
GENERATING STRUCTURE
PERMIAN 6500 FEET
PECOS CO., TEXAS
ELEMENTAL COMPOSITION
WT. %
ATOM
80% CARBON 6.7
8% HYDROGEN 8.0
10% OXYGEN .6
2% NITROGEN .1
H H CH3
HO OH
CH3 (CH2)10-CH3
H H
H CH2-CH3
B. HYPOTHETICAL GAS
GENERATING STRUCTURE
MIOCENE 14,500 FEET
CAMERON PH., LOUISIANA
ELEMENTAL COMPOSITION
WT. %
ATOM
80% CARBON 6.7
5% HYDROGEN 5.0
13% OXYGEN .8
2% NITROGEN .1
OH H OH CH3
H OH
CH3
H
O H
H H CH2 H
CH3

Fig. 1. Structure of marine kerogen

The distribution of marine kerogen in recent sediments is shown in Table I (8) and Table II (6). The kerogen content in both cases was much higher than the hydrocarbon or asphalt content. The hydrocarbon content of basin sediments off Southern California is shown in Table III (10).

Table I. Distribution of Hydrocarbons and Associated Organic Matter in Recent and Ancient Non-Reservoir Sediments (ppm) (Ref. 8)

Recent Sediments	No. Samples	Hydrocarbon	Asphalt	Kerogen
Mediterranean Sea	1	29	461	9000
Gulf of Mexico	10	12-63	113-790	3800-9100
Gulf of Batabano, Arba	10	15-85	136-1023	2900-34,900
Orinoco Delta, Venezuela	10	27-110	283-1355	7500-14,700
Lake Maracaibo, Venezuela	8	24-116	266-1600	13,700-41,500
Ancient Sediments				
Shales	791	300	600	20,100
Carbonates	281	340	400	2,160

Table II. Distribution of Organic Matter in Recent Sediments (BBL/ACRE)(Ref.6)

Sample	Hydrocarbon	Asphalt	Kerogen
Gulf Coast	1.3	3.0	115
California Coast	4.2	15	420

Table III. Hydrocarbons in Basin Sediments off Southern California (Ref. 10)
(Dry Sediments)

Basin and Depth In Sediment (CM)	C(%)	Organic Extract (G/150G)	Paraffin-Naphthenes (ppm)	Aromatics (ppm)	Asphaltenes (ppm)
Santa Barbara Basin					
0-16	3.75	0.301	132	247	1200
32-60	3.24	0.239	100	172	1460
100-152	----	-----	91	121	----
248-274	2.40	0.175	72	61	1420
418-478	2.06	0.166	81	119	1230
Santa Monica Basin					
0-40	----	0.195	51	99	1210
San Diego Trough					
5-30	----	0.220	64	126	1360
Santa Catalina Basin					
0-60	3.63	0.141	10	23	746
60-148	3.66	0.138	77	69	668
206-250	3.50	0.092	11	18	485
294-340	1.63	0.048	9	22	196
Santa Cruz Basin					
24-36	3.76	0.188	36	64	277
274-304	2.91	0.152	29	254	272
396-442	3.36	0.149	38	125	408
Tanner Basin					
3-38	-----	0.144	129	38	483
64-152	2.91	0.152	29	254	272
305-331	----	0.094	14	25	595
San Clemente Basin					
0-5	----	0.065	21	22	233
Long Basin					
0-30	----	0,097	20.4	34	398
Soledad Valley					
0-5	----	0.76	501	296	----

Table IV. Bitumen Components in Recent Bering Sea Sediments (Ref. 11)

Type of Sediment	Bitumen Components (%)			Content of Oils
	Asphaltenes	Resins	Oils	(% of Dry Sediment)
Sands	64	23	11	0.004
Coarse Silts	62	25	10	0.005
Fine Salt Muds	65	24	11	0.006
Silt-Day Muds	62	27	11	0.008
Day Muds	64	26	10	0.014

Marine bitumen originates from the lipid fraction of phytoplankton. It can also be derived from kerogen and represents the greatest potential source of oil formation. The average composition of bitumens in the upper layer of marine sediment is 65% asphaltene, 25% resins, and 10% oil (Table IV) (11). Despite some variations, group compositions of marine bitumens are fairly uniform. The mean bituminosity of marine sediments is 0.006% (Figs. 2, 3) (11). Organic carbon associated with bitumen in the marine sediments averages about 5%. In contrast to kerogens, the bitumens are soluble in organic solvents (Table V).

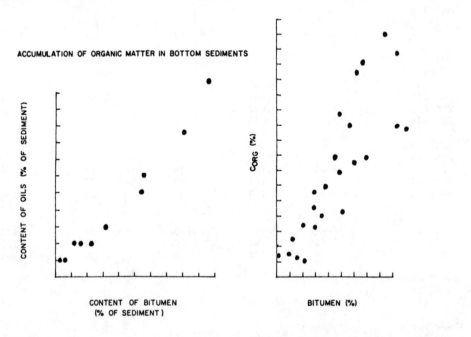

Fig. 2. Relation of bitumen to oil

Fig. 3. Relation of organic carbon to bitumen

Table V. Extraction of Miscellaneous Recent Sediment Samples of Dry Sediment

	Organic Matter Extracted (ppm)	Chromatographic Anglysis of Extract		
		% Prarffin- Naphthene	% Aromatic	Total Hpdrocarbons (ppm)
Mississippi Delta Samples				
BC 236-52	2,500	1.6	2.7	110
PL 242-52	326	17.6	10.5	92
GM 311-52	457	4.8	2.2	32
MP 321-52	362	1.3	6.1	27
Louisiana Coast Samples				
Marsh Samples	490	0.8	1.1	9
Mud-flat Deposit	1,040	2.7	2.3	52
California Coast Sampels				
Estuary Pound	7,600	6.6	3.9	800

Bituminous shale, natural seeps and tar sands contain relatively high quantities
of bitumen. Bituminous shales are closely-spaced bituminous laminae averaging
about 20 microns in thickness. The organic carbon content ranges from 5 to 15%
(12). In basin floor sediments there are both bitumen (soluble) and kerogen
(insoluble) coexisting (Table VI). Natural seeps and tar sands found along the
shallow continental shelf region average 15% in bitumen. Along the California
coastline, two rich deposits, the Monterey Shale and Edna Tar Sands have been
located. In contrast, natural seeps including tar balls along the Santa Barbara
Basin have been only recently studied (Fig. 4) (13).

Edna is the largest tar sand deposit in California. The structural basin is in
the Coastal Range district near San Luis Obispo. It is the outcrop of the generally massive, diatomaceous shale of the Monterey Formation. Another tar sand
deposit in California is the Sisquoc deposits in Santa Barbara County near Salmon
Hills. It seems the Sisquoc deposits are folded and faulted inclusions of the
underlying Monterey shale stretching to the continental margins. The oil beds of
Miocene, Monterey and Voquerous Formations such as the Santa Cruz Asphalt deposits
are believed to be the result of indurated and jointed pools under faulting and
pressure.

DISCUSSION

Due to the similarities in genesis, physical, and chemical properties between the
marine and oil shale kerogens and between the marine bitumen and tar sands, both
marine kerogen and bitumen have undergone serious consideration as alternative
energy sources.

Using the formula $C_{215}H_{330}O_{12}N_5S$ for the oil shale kerogen molecule (14), calculation of the gross heating value may be obtained using Dulong's formula:

Table VI. Some Characteristics of Basin Floor Sediment (Ref. 10)

		Soluble Residue Median Diameter (μ)	Trask Sorting Coeffi.	Organic Matter %(17xN)	Hydrocarbons (ppm)	Insoluble Residue Median Diameter (μ)	Trask Sorting Coeffi.
1	Los Angeles	14.0	3.9	2.6			
2	Santa Barbara	4.5	3.4	5.8	240	6.7	2.4
	Santa Monica	8.5	3.2	6.5	150	6.1	3.4
	San Pedro	7.5	3.6	7.3		12.5	2.6
	San Diego	5.6	4.6	8.5	200	8.2	2.4
3	Santa Cruz	3.9	3.6	8.5	180	8.5	2.2
	Santa Catalina	4.0	3.7	8.6	60	8.9	2.4
	San Clemente	4.0	4.2	5.7	43	10.9	2.4
4	San Nicolas	3.7	3.9	8.1		6.3	2.7
	East Cortes	3.7	3.2	6.4		4.7	2.5
	No Name	4.6	2.1	4.6		7.9	2.3
5	Tanner	4.8	3.4	11.7	98	5.6	2.8
	West Cortes	6.4	3.7	6.4		7.7	2.4
	Long	3.1	3.2	5.0	54	6.7	2.4
6	Continental Slope	8.0	4.4	4.1		6.7	2.6
7	Deep Sea	1.8	2.7	2.1		2.9	2.8

O BITUMINOUS SHALES
● TAR SANDS OR ASPHALTS
▨ NATURAL SEEPS

Fig. 4. Natural resources in continental margins of California

$$Btu/lb = (14{,}600)(\text{fraction C}) + (61{,}000)(\text{fraction H} - 1/8 \text{ fraction O}) + (4{,}000)(\text{fraction S})$$

where 14,600, 61,000 and 4,000 represent the approximate amount of heat produced by combustion of one pound of C, H_2, and S, respectively. Fractions of C, H, O, and S are expressed as weight percentage. The approximate gross heating value of pure kerogen is about 176,000 Btu/lb(14).

It is possible to apply this value to marine kerogen. For example, the average concentration of kerogen in recent sediments in the Gulf of Mexico is 6,400 ppm. The amount of heat generated by kerogen from there may be equivalent to 20 cubic miles of coal or one hundred billion tons of fuel oil. During a heating period of 10 minutes at 500°C, 60% of the oil shale kerogen is converted to oil and 23% to gas. Oil shale type marine kerogen would produce a similar quantity oil and gas, while coal type marine kerogen would produce a higher percentage of gas than oil. The average composition of Athabasca bitumen is 19% asphaltene, 32% resins, and 49% oil (14). This deviation from the composition of bitumen in the Bering Sea sediments could be accounted for by the differences in geobiological environments since both asphaltene and resin can be converted to oil through microbial modification. Such microbiological modification is well substantiated for the Athabasca sediments (17). The heating value of Athabasca bitumen averages 17,700 Btu/lb (14). This value should also be anticipated for bitumens of continental margins.

PROPOSED RECOVERY METHODS

1. Marine sediments are dredged along the continental shelf region and piled up for deposition at an assigned spoil bank (Fig. 5). A bound dike is formed by fencing with an asphalt-sand emulsion (19). Offshore or beach sand and gravel can be used to cover the top of deposited sediments. Under anaerobic conditions, heat from sunlight will set in and initiate fermentation (Fig. 6). The technology is similar to that of landfill gas production (18). The hydrolyzable carbon in organic matter will produce methane. The produced gas can then be delivered by wells to inland pipelines.

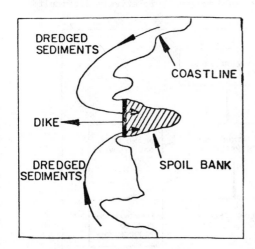 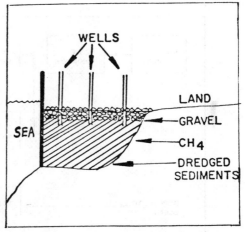

Fig. 5. Dredged spoil site   Fig. 6. In situ fermentation

2. A marine operation is constructed which includes several separating cells along the coastal line. The dredged natural seeps, bituminous shales, or tar sands can be treated with cold water extraction method for recovery of bitumens. Cold water extraction processing has been employed in the recovery of bitumen from tar sands by using soda ash, wetting agent and cutback light oil at 21°C to 26°C. This method can recover up to 95% of the bitumen in oil sands (15).

3. Enzyme treatment is another method of extracting bitumen or kerogen. Enzymatic reactions can separate organic material from inorganics by breaking the linkages between organic carbon and mineral particles. The organic matter is then treated by a flotation method and pressure so that the oil is recovered at the

62

top of the setting cell. It is relatively difficult to cleave the bond between marine kerogen and clays; hence mild oxidation of kerogen to bitumen or humic acid may be necessary prior to enzyme treatment. In addition, ozone would be the most desirable oxidizing agent (16).

FEASIBILITY

Marine sediments are naturally-abundant alternative energy resources. Kerogen and bitumen are the major organic substances under consideration here. Marine kerogens comprise the major portion of organic matter in sediment. These kerogens can be converted to gas, oil, bitumen or resin with the amount of gas or oil produced depending upon the type of kerogen. Bitumen, however, is converted primarily to oil.

According to the proposed methods discussed, whose outline is summarized in Fig. 7, dikes would be formed by disposing dredged sediments at spoil banks and fenced using an asphalt-based material. The fermentation reaction, under the anaerobic conditions produced by covering with gravel and sand, would yield methane for inland or coastal zone use. The recovery of bitumen from bituminous shales, natural seeps, or tar sands would be accomplished through cold water extraction. Enzymatic reactions followed by the flotation process would recover the oil or gas from marine kerogen. However, in situ ozone oxidation is recommended prior to enzymatic treatment, otherwise the recovery of kerogen is relatively difficult.

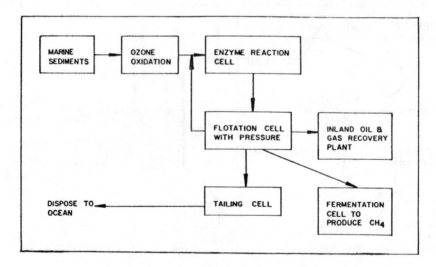

Fig. 7. Flow diagram of energy recovery of marine sediments

ACKNOWLEDGEMENT

Partial support from A.G. A. BR-48-12, ERDA-AEC E(29-2)-3619 and NOAA-Sea Grant R/RD-2 is acknowledged.

REFERENCES

1. B. A. Skopintesev, 1950, "Organic Matter in Natural Water (Water Humus)," Tr. Geol. Inst. Akad. Nauk S.S.S.R., Vol. 17, No. 29.

2. T. F. Yen, 1972, "Terrestrial and Extraterrestrial Stable Organic Molecules," in Chemistry in Space Research, Elsevier Scientific Publishers, New York, pp. 105-153.

3. T. F. Yen and J. I. S. Tang, 1977, "Chemical Aspects of Marine Sediments," in Chemistry of Marine Sediments, Ann Arbor Science Publishers, Chap. 1, pp. 1-38.

4. R. D. McIver, 1967, "Composition of Kerogen: Clue to Its Role in the Origin of Petroleum," 7th World Petroleum Congress Proc., Vol. 2, pp. 25-36.

5. J. P. Forsman and J. M. Hunt, 1958, "Insoluble Organic Matter (kerogen) in Sedimentary Rocks of Marine Origin," in Habitat of Oil, (L. G. Wecks, ed.), Am. Assoc. of Petrol. Geol., pp. 747-778.

6. J. M. Hunt and G. W. Jamieson, 1956, "Oil and Organic Matter in Sources Rocks of Petroleum," Bull. Am. Assoc. Petrol. Geol., Vol. 40, pp. 477-488.

7. R. E. Laphante, 1971, "Hydrocarbon Generation in Gulf Coast Tertiary Sediments," Am.Assoc, Petrol. Geol. Bull., Vol. 58, pp. 1281-1289.

8. J. M. Hunt, 1961, "Distribution of Hydrocarbons in Sedimentary Rocks," Geochim. Cosmochim. Acta, Vol. 22, pp. 37-49.

9. J. M. Smith, 1954, "Conversion Constants for Mahogany-Zone Oil Shale," Am. Assoc. Petrol. Geol. Bull., Vol. 50, pp. 167-170.

10. K. O. Emery, 1960, The Sea Off Southern California, John Wiley and Sons, Inc., New York.

11. O. K. Bordovskiy, 1964, "Accumulation and Transformation of Organic Substances in Marine Sediments," marine Geol., Vol. 3, p. 83.

12. A. Hallam, 1967, "The Depth Significance of Shales with Bituminous Laminae," Marine Geol., Vol. 5, pp. 481-493.

13. T. C. Henyey and T. F. Yen, 1975-1977, "Oil and Tar Seeps Off Southern California," University of Southern California Sea Grant Research Project.

14. Synthetic Fuels Data Handbook, 1975, Cameron Engineers, Inc.

15. L. E. Kjinghenzian, 1951, "Cold Water Method of Separation of Bitumen from Alberta Bituminous Sand," in Proceedings of the 1st Athabasca OIl Sands Conference, Ings Printer, Edmonton, Alberta, Canada, pp. 185-199.

16. T. F. Yen, 1974, "Feasibility Studies of Biochemical Production of Oil Shale Kerogen," A Preliminary Report, NSF-RANN.

17, T. F. Yen, 1975, "Genesis and Degradation of Petroleum Hydrocarbons in Marine Environments," Marine Chemistry in the Coastal Environment, pp. 231-266.

18. D. C. Argenstein, D. C. Wise, R. C. Wentworth and C. L. Cooney, 1976, "Fuel Gas Recovery from Controlled Land-Filling of Municipal Wastes," Resources Recovery and Conservation, Vol. 2.

19. C. J. Krom, Private Communication (Koninklijke/Shell Laboratorium, Amsterdam).

7

URBAN AND FISH PROCESSING WASTES IN THE MARINE ENVIRONMENT:
A CASE OF WASTED ENERGY AT TERMINAL ISLAND, CALIFORNIA

by

Dorothy F. Soule, Mikihiko Oguri, and John D. Soule
Harbors Environmental Projects

INTRODUCTION

The coastal environment has long served as a place for man to dispose of urban domestic and industrial wastes that can include a wide range of substances ranging from non-toxic, biodegradable nutrient wastes to oxygen-depleting or toxic chemicals. It is now also recognized that the coastal waters along the continental shelf are much more productive than deep sea waters, largely due to nutrients of terrestrial origin. The rich and diverse coastal eco-systems once depended primarily upon natural run-off and sometimes upon upwelling, which can bring nutrient materials that have drifted from shallow coastal waters into deep canyons back up to productive surface waters.

Where man has progressively reduced or eliminated marshlands, which act as nutrient sinks and hence as marine nursery grounds, and has dammed or concretized river channels, urban wastes may offer almost the only major regular source of nutrient energy input to coastal waters. Thus the efforts of enforcement agencies to mandate secondary waste treatment for all wastes, toxic or non-toxic, may ultimately deprive the marine environment of nutrient energy upon which it has come to depend.

Problems that have been associated with domestic and industrial waste disposal are derived from the lack of limitations on the sewering of toxic substances, on the mixing of non-point source storm run-off with normal domestic waste loads in sewerage, and on disregard for the limit in capacity of receiving waters to recycle wastes without deteriorating environmental quality. Most natural products are bio-degradable; that is, microbial populations are capable of breaking down (digesting) the larger molecules and releasing simpler forms of carbon, hydrogen, nitrogen, and phosphate to be recycled through the food web of marine and freshwater habitats. Many small invertebrates feed almost exclusively by filtering out bacteria and other microbials; these invertebrates are then fed upon by larger invertebrates and vertebrates. There is also considerable evidence that direct uptake of larger organic molecules occurs without passing through a food chain. These energy pathways could be utilized for creating inshore or onshore mariculture of plants and animals.

On the other hand, the EPA-mandated secondary waste treatment creates a microbial digestion within the treatment plant. The large populations of bacteria and other microbes (the floc) which grow on the nutrient wastes become a new problem as they die off within days, forming masses of sludge that also must be disposed of -- on land according to Environmental Protection Agency edicts. Toxic wastes, heavy metals, pesticides and the like are often not biodegradable and are trapped either in the solid waste sludge or are dissolved in the effluent that goes into rivers, lakes or the sea as supposedly "clean" water.

The many problems associated with large urban all-purpose systems are related to the lack of information on the sources and the composition of the waste stream. Human wastes of themselves are not toxic; they are biodegradable and have been used historically around the world as fertilizer. The danger with untreated human waste usage lies in the transmission of disease if contaminated wastes come in contact with people working the land or enter the drinking water supply.

Problems of waste disposal also arise where waste loads entering waters exceed the capacity of the localized area of the water body to maintain oxygen levels suitable to survival of organisms. Waters must supply both the oxygen used by bacteria in metabolic breakdown and recycling of the organic matter (Biochemical Oxygen Demand), and the oxygen used up by oxidation of chemicals (Chemical Oxygen Demand) in the wastes, as well as provide dissolved oxygen for respiration and metabolism of organisms. Similar problems of oxygen capacity occur with food processing plant wastes. Such wastes generally do not contain toxic materials and would be biodegradable if the wastes could be delivered to waters with sufficient circulation and oxygenation to recycle the nutrients without depleting dissolved oxygen below levels necessary for survival of organisms such as fish.

Oxygen budgets can be calculated and new techniques of computer modeling offer methods for managing waste loads to maintain adequate oxygen levels while still retaining the nutrient input to the marine environment. Rather than using nutrients only to feed bacteria in secondary waste treatment plants and create sludge, nutrients could be returned to the environment or, even more important, could be used for increased production in inshore waters or for mariculture, either in onshore facilities or in semi-contained waters. Proteins, amino acids, carbohydrates and fats are present in fish cannery wastes and in domestic wastes to varying extents; these can be used directly or indirectly by marine organisms. In contrast, reduced levels and differing forms of nutrients in secondary waste effluents are usable only by phytoplankton and some protists, which may produce "blooms" and certainly do not support the same diverse ecosystem that primary (screened but undigested) wastes support.

INVESTIGATIONS

Studies by Harbors Environmental Projects at the University of Southern California have been carried out intensively in the Los Angeles-Long Beach Harbors of San Pedro Bay, California since 1970. The harbors are man-made, developed from the estuarine system of the Los Angeles and San Gabriel Rivers. Because southern California has a winter season with an average of 14 inches (36 cm) of rain (with extremes from 4" to 36") and a dry summer, there is no continuous riverine flow of water, and salinity gradients are transitory. Figure 1 shows the change from the shallow water, sand bar, and mud flat geography of the harbor in 1872 with the present configuration superimposed. Dredging and landfill have created main channel depths of about 35-45 feet (10-14m) on the west side of the harbors and subsidence due to oil extraction has created some 60 foot (18 m) depths in the eastern harbor area.

Fish canneries have been located in the harbor since the turn of the century and until 1977 three canneries were located on Terminal Island, along with a small primary treatment sewage plant. With the growth of the oil industry in southern California, extraction and refinery activity increased in the harbor in the 1920-1960's period. Dumping of all sorts of wastes into the harbor created virtually anoxic waters in the innermost slips and very poor quality for all but the outer harbor. Construction of the Federal (middle) breakwater and the Navy mole during World War II and completion of Pier J in Long Beach in 1970 reduced water circulation even further.

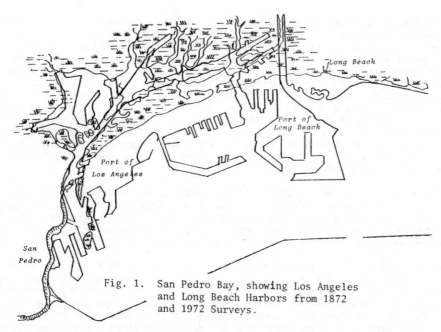

Fig. 1. San Pedro Bay, showing Los Angeles
and Long Beach Harbors from 1872
and 1972 Surveys.

With the coming of environmental legislation (The California Environmental Quality
Act and the National Environmental Policy Act), control of toxic refinery wastes
and other industrial effluents produced dramatic improvements in the water quality
and biology of the harbor. Attention then turned to the canneries, which had for
years dumped bones, scales and guts into the harbor and fishing boats had flushed
their holds. Economics first dictated that all large particulate wastes be recovered for use in fish meal or pet food, but large waste loads of water carrying
oils and both dissolved and suspended organic solids created heavy oxygen demand
on the waters. Peak cannery loading occurred in the fall when wet fish (anchovy)
were processed at natural seasonal times when cooler nights or strong desert winds
caused overturn of warmer bottom waters, resuspending anaerobic sediments and sti-
mulating red tide blooms. The combination of natural and industrial oxygen demand
sometimes produced anoxic conditions.

In spite of the periodic low-oxygen episodes, research by the Harbors Environmental Projects team showed that the harbor had a rich and diverse soft-bottom community. Furthermore, in working with the canners they found that baseline data
could be used to develop a computer model of the capacity of the receiving waters
to assimilate these organic wastes. Canners began to limit processing loads in
relation to projected conditions for harbor waters, and for more than two years
succeeded in avoiding any anoxic conditions such as those that led to fish kills
in previous years. It was under these conditions that the value of the cannery
and primary treatment wastes in the harbor became apparent.

In an energy-poor environment it has now become important to develop alternative
methods of waste management and new regulatory concepts, rather than destroy
energy-rich nutrients before they enter the sea. Our society can no longer afford
to destroy nutrients in one place only to import them at great expense from another.

The following statement was presented to the California Water Pollution Control
Association in 1978 by the senior author, summarizing the background of regulatory agency action and the results of USC investigations:

68

"In October 1976 Harbors Environmental Projects (HEP) of the Institute for Marine
and Coastal Studies at the University of Southern California, presented a report
of several years of investigations on the effects of waste effluents on the outer
Los Angeles Harbor ecosystem. This report was incorporated as a part of the Draft
Environmental Impact Report for the City of Los Angeles Terminal Island Treatment
Plant (TITP). Because of numerous requests this study was published with some
additional data in December 1976 as Volume 12 of our Marine Studies of San Pedro
Bay, California (Soule and Oguri, 1976). These studies were funded in an unusual
cooperative effort by five seafood processors (only three of which are located in
the harbor), by the City of Los Angeles Department of Public Works Bureau of Engineering and by the Federal Sea Grant Program (NOAA, Department of Commerce), as
well as with funds from USC's HEP, the Institute of Marine and Coastal Studies
and the Allan Hancock Foundation.

"The receiving water area of TITP and the canneries had been under study since
1970 by HEP, under a program funded in part by the Southern California Gas Company,
the Port of Los Angeles, the Port of Long Beach, the Corps of Engineers, the Tuna
Research Foundation, StarKist Foods and other local industries of agencies. Thus
these studies cannot be considered as vested interests of any of the participants.

"Findings then current were used as a basis for input on the California State
Water Quality Control Policy for Enclosed Bays and Estuaries of California, May
1974. This clearly states in Chapter I, Section A that '...discharge of municipal wastewaters and industrial process waters to enclosed bays and estuaries,
other than the San Francisco Bay-Delta system, shall be phased out at the earliest
practicable date.' It then states that '...exceptions to this provision may be
granted by a Regional Board only when the Regional Board finds that the wastewater
in question would consistently be treated in such a manner that it would enhance
the quality of receiving waters above that which would occur in the absence of the
discharge...' (author underline).

"Section B goes on to discuss the San Francisco-Delta problems and takes note of
the evidence suggesting that biological degradation is due to long-term exposure
to toxicants that have previously been discharged into the system. However, no
mention is made of a similar, obvious situation in Los Angeles Harbor, which was
long used as an industrial waste dumping ground by oil and chemical industries,
shipyards and others.

"The studies undertaken by USC were not limited to the standard water quality enforcement measurements. Public agencies already provide or require submission of
much of these data. Rather, our investigations were of several sorts, with the
following scope:

1) To develop from the existing data and literature, from our own research, and
   from that of others, evidence of areas in the harbor that could be considered
   enhanced, and those that could be considered impacted;

2) to compare these areas with other similar ones in southern California in terms
   of ecological richness and impacts;

3) to identify, where possible, the reasons for the enrichment or impacts in the
   harbor area;

4) to suggest mitigating measures, where feasible;

5) and to develop methodologies for managing the input of nutrients to maintain
   an ecologically valuable (bioenhanced) area in a homeostatic system, while
   maintaining water quality standards.

6) While cost-effectiveness was not a concern of the funded studies, as citizens we cannot avoid being concerned at the obvious economic and social impacts of proposed enforcement policies which may in fact be unnecessary and also result in degradation of the environment rather than benefitting it."

A summary of the research (Soule and Oguri, 1976) was presented, as follows:

"The receiving waters for fish cannery wastes in outer Los Angeles Harbor have been studied by Harbors Environmental Projects of the University of Southern California, since 1970. During that period, field investigations have been made of physical and biological parameters on a monthly basis, with specialized studies being carried out biweekly, weekly and daily during a portion of that time.

"Physical conditions surveyed include circulation and flushing, temperature, dissolved oxygen, pH, salinity, turbidity, sediment character, pollutants, BOD and nutrients. Biological parameters include microbiology, phytoplankton productivity, zooplankton, benthic and water column invertebrates, fish and birds.

"Laboratory studies have been carried out on bioassays, reproduction and growth, stress, toxicity, and food web relationships.

"Mathematical modeling studies use the baseline data to relate the parameters to one another and work toward projection of organic loading in relation to assimilation capacity of the receiving waters.

"The following statements summarize the information and conclusions derived from these investigations:

1) The field studies indicated that the existing state of the harbor was healthy. Rich and diverse biotic elements were supported by the extent of the environmental regime. Episodes of stress, which occurred in earlier years, as indicated by reduced levels of dissolved oxygen, had not been noted since the canneries instituted improved waste management procedures.

2) Bioenhancement (the enhancement of the biological quality of receiving waters) had not occurred in outer Los Angeles Harbor, due at least in part to the presence of natural waste effluents from the primary treatment of the Terminal Island Treatment Plant and the canneries.

3) Bioenhancement has been evaluated in terms of numbers of organisms and species diversity of plankton, benthic organisms, and standing crop of fish, as well as in biomass and a number of other factors detailed in the research reports.

4) The fish populations were higher in the outer harbor than in any similar coastal area in southern California. The harbor has been essential nursery grounds for the 0-1 year age class of anchovy and for other fish species.

5) Under conditions that existed until the fall of 1977, there was a small zone within approximately 200 feet of the outfalls where numbers of species were low. Adjacent to this zone was a zone of enrichment which extended through most of the outer harbor. Beyond that, conditions were of average coastal populations. The regulation of waste loading and control of pollutants in the past six-year period had brought the harbor ecosystem from a depauperate biota to a moderately rich one in the immediate outfalls zone, with a very rich biota in the adjacent outer harbor area. This was due in large measure to enforcement and control of toxic industrial and refinery wastes and to management of cannery waste loads.

6) There was a net bioenhancement over and above those conditions which would occur in the absence of the existing natural waste discharges.

7) It was postulated by HEP that cessation of all effluents would probably cause a gradual or accelerated reduction in the biota and ecosystem. Such phenomena have been documented in the United States and elsewhere.

8) The organic load from the cannery wastes put a high Biological oxygen Demand (BOD) on receiving waters. At the same time it is important to emphasize that BOD represents high nutrient input to an ecosystem, provided that sufficient dissolved oxygen is available to prevent reduced water quality. Many harbor organisms are detrital feeders, dependent upon filtering or ingesting bacteria.

9) A more limited biota, tolerant to the effluents, was found in a relatively small area near the discharge points which were in 3-4 meters of water. Harbor organisms more sensitive to the effects of the effluent were not usually found there and on laboratory testing were unable to survive in high concentrations of the effluent. This is due to a number of factors: fresh water, chlorinated water, BOD, trace metals and perhaps industrial chemicals. The area has also served as a dump site and ad hoc ship scrapping area, which does not enhance water quality.

10) Management strategies can be developed to predict generally the amount of loading possible under various environmental conditions. Mathematical model studies of the harbor based on the data collected, suggested that the assimilation capacity of the receiving waters was not being exceeded by the organic load discharged in these waters. The model studies were being further developed to reflect short-term stress and change, so that they may be used ultimately for other regions, outside the United States if not within it.

"It is unfortunate that while this cooperative, largely local effort was making good progress toward managing the effluents in outer Los Angeles Harbor, regulatory agencies were inexorably moving toward enforcement of traditional waste treatment ... The harbor may have been serving as a giant oxidation pond, but the diverse bacteria that break down proteins, carbohydrates, fats, oil and amino acids in turn were immediately eaten by the swarms of tiny crustaceans and other zooplankton and by numerous detritus-feeding benthic worms. The worms were consumed in huge quantities by obligate or omnivorous bottom feeding fish, molluscs and crustaceans. Nutrients were recycled by phytoplankton blooms on which zooplankton and juvenile fish, including anchovy larvae, depend.

"In justice to the regulatory agencies, traditional sanitary engineering practices dictated secondary waste treatment for domestic waste effluents. The traditional methods of dealing with high BOD wastes are directed toward making river and lake water suitable for drinking water for the next city downstream. The wastes are not supposed to feed anything, as a priority over being potable, -- which is, of course, contrary to what we need in our coastal ecosystems. BOD is usually considered to be A BAD THING, to be eliminated wherever possible. The fact that BOD is removed the waters would have little nutrient value left, has been largely ignored for ocean disposal systems."

DISCUSSION

The premise that waste effluents from domestic sewage (Goldman and Ryther, 1975; Howell, 1977; Woodwell, 1977) and from food processing plants can enhance the ecosystems of marine waters is not unique to the Terminal Island effluents in outer Los Angeles Harbor. However, documentation of the effects of cannery waste fields

is scarce, while sewage outfalls have been more extensively studied. Empirical
evidence of enhancement is often observed-- by the anglers who fished with un-
baited gang hooks from the shore beside the outfall pipes at Terminal Island, by
the sport fishing party boats that regularly frequent the local ocean sewer outfalls, and by the fishing birds such as pelicans that dive repeatedly into the
effluent areas to capture smaller fish attracted to the food source.

Locally in southern California, field studies of receiving waters have been carried
on at Hyperion outfall in Santa Monica Bay (Los Angeles City), at Whites Point outfall off Palos Verdes Peninsula (Los Angeles County), at the Orange County outfall,
and at the Avalon outfall on Santa Catalina Island. The Allan Hancock Foundation
(University of Southern California) first studied the Hyperion outfall in 1946-47,
and covered the entire coast of southern California beyond about the 100 ft contour where waste fields occur (AHF, 1965). The Southern California Coastal Water
Research Project was inititated in 1969 by the Cities of Los Angeles and San Diego,
the Sanitation Districts of Los Angeles and Orange County and by Ventura County,
to study the receiving waters and the impacts of waste input. A number of specialized studies and reports have originated from that group (SCCWRP, 1974, 1975, 1976,
1977). Bascom (1974, 1976), director of SCCWRP, wrote on waste disposal in the
ocean and noted that a zone of enhancement lies outside the immediate zone of impact of the discharge. D.J. Reish and his associates (pers. comm.) and Emerson
(1974) reported on the biostimulatory effects of certain dilute concentrations of
wastes on polychaetes.

The degree to which field conditions in the Los Angeles Harbor receiving waters
for the Terminal Island cannery wastes have been documented is perhaps unique.
Probably no other cannery waste fields have been as extensively studied. The results of these and other studies have helped to form the basis of the conclusion
that the natural wastes provide enhancement of the local ecosystem. The occurrence
of a large circulation gyre in the outer harbor was described from a drogue study
and confirmed by current meter studies (Soule and Oguri, 1972; Robinson and Porath,
1974). The Army Engineer Waterways Experiment Station physical model in Vicksburg,
Mississippi ultimately duplicated this gyre (McAnally, 1975). Concentrations of
trace metals and other pollutants in the harbor and in the San Pedro channel area
were mapped for the first time (Chen and Lu, 1974), while another HFP study mapped
biomass distribution of benthic organisms in the outer harbor, measured for benthic
recolonization potential and carried out bioassays showing the effects of resus-
pension of sediment (dredging or stirring).

Figure 2 shows the station pattern for the 1973 and 1974 study for the Army Corps
of Engineers (AHF, 1976) under which a data bank was created permitting multivari-
ate computer analysis of the interactions of parameters.

Circulation patterns of the harbor, shown in the model studies, are presented in
Figures 3 and 4 (McAnally, 1975). These can be compared with the only previously
published illustration of harbor circulation before the extensive Pier J landfill
(Fig. 5, adapted from Reish, 1959). Flushing was quite different in the harbor,
and the gyre was not apparent. Circulation now appears to be strongly wind-driven
in the outer harbor because it is relatively shallow. Water quality parameters
for the harbors have been reported also (Allan Hancock Foundation, 1976; Soule
and Oguri, 1974,1976).

Benthic Animal Life

During the HEP study in 1973 and 1974, benthic samples were taken throughout the
harbor on a quarterly basis. The diversity of species present and the numbers of

72

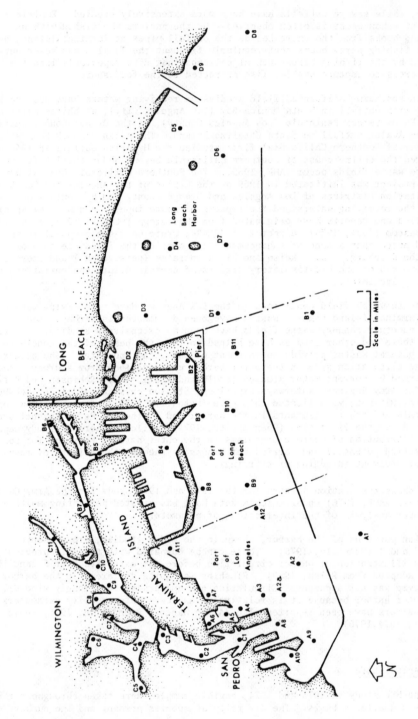

Figure 2. Station pattern of sampling in Los Angeles-Long Beach Harbors by Harbors Environmental Projects, 1973 and 1974.

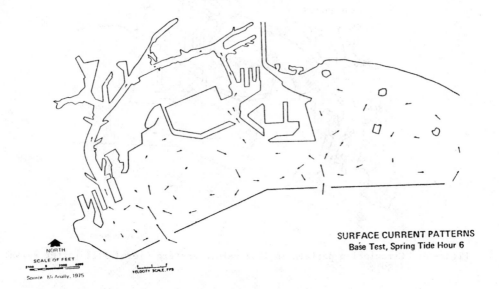

Figure 3.   Circulation pattern on rising tide, in physical model.

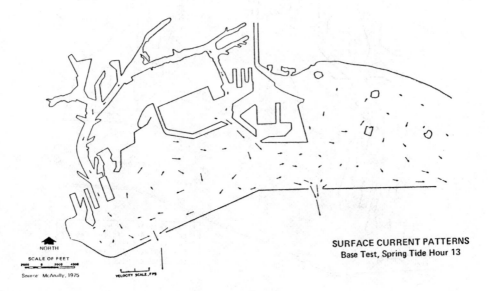

Fig. 4.   Circulation pattern on falling tide, in physical model.

74

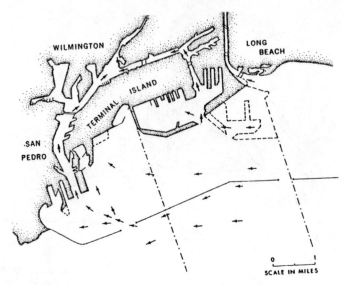

Figure 5. Circulation pattern in 1954 before eastern Pier J fill (after Reish, 1959).

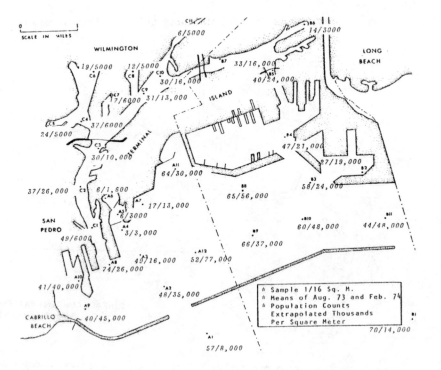

Figure 6. Numbers of benthic species/numbers of individuals per square meter.

75

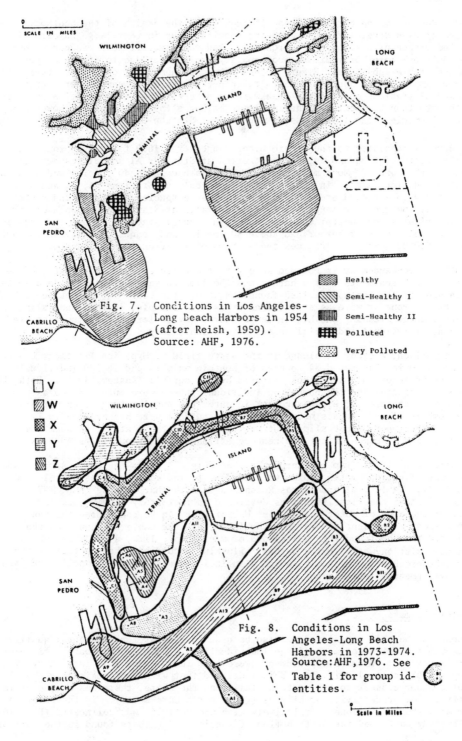

Fig. 7. Conditions in Los Angeles-
Long Beach Harbors in 1954
(after Reish, 1959).
Source: AHF, 1976.

Healthy
Semi-Healthy I
Semi-Healthy II
Polluted
Very Polluted

V
W
X
Y
Z

Fig. 8. Conditions in Los
Angeles-Long Beach
Harbors in 1973-1974.
Source:AHF,1976. See
Table 1 for group id-
entities.

Scale in Miles

76

individuals combine to give a good indication of the health of the benthos. Reduced species diversity, even when the species occur in very large numbers, is
an indication of a stressed environment. Reduced diversity and reduced numbers
(production) together indicate poor quality environment due to stress or perhaps
to unsuitable substrate. Figure 6 shows the mean numbers from representative boxcorer samples, with species/numbers taken per square meter sample per station. The
A1 and B1 stations outside the breakwater can be taken as a comparative standard,
with more than 50 species. The innermost harbor slips and channels show fewer
than 50 species; the lowest was six species at Station C11.

Numbers extrapolated to the square meter ranged from 1600 to 6400 in inner slips,
while counts of 8000 and 14,000/m^2 occurred at Stations A1 and B1 respectively,
outside the harbor. Most main channel stations had 30 to 40 species and 6400 to
26,000 individuals/m^2. In the outer harbor, only at Stations A4, A5 and A6 associated with enclosed areas at Fish Harbor, were species diversity and numbers
both low. Between the cannery and sewer outfalls at Station A7, counts of numbers
of individuals were better than several main channel stations. Most stations in
the rest of the outer harbor had above 50 species and counts of individuals ranged
up to 76,000, some 7 to 10 times the numbers of individuals at A1 and B1.

Benthic bioenhancement is not due solely to the waste effluent nutrients, of course,
for finer sediments in the harbor and the limited wave action provide a suitable habitat for the specialized fauna. However, the counts would not be so high
if the environment were as stressed as it is, for example, in the inner slips,
where reduced flushing and higher temperatures may deplete the bottom dissolved
oxygen, or residual industrial wastes may be inhibitory.

In comparison, a study of sludge in the waste field of Hyperion Treatment Plant
in Santa Monica Bay found 24 species of benthic animals and 16,000 individuals
per square meter in the benthos affected by the outfall (Bascom, 1976). Both the
number of species and the number of individuals there were much reduced as compared with the area of bioenhancement in outer Los Angeles-Long Beach Harbors.
The numbers of individuals near the Hyperion outfall were higher than in Fish Harbor and the inner harbor slips, but were similar to those at A7, between the Terminal Island Treatment Plant and cannery effluents in the outer Los Angeles Harbor.

Reish's data (1959) from the harbor in 1954 (Fig. 7) shows the areas sampled and
the classification of the sediments according to indicator species he selected
(Table 1). Table 1 also compares Reish's data with that of HEP (AHF, 1976), and
Figure 8 shows the classifications developed in that study. Reish sampled only a
part of the area of outer Los Angeles Harbor that can be considered the zone of
enhancement at present. However, the figure and the table do show that the most
polluted zone he found then was devoid of macroscopic life, a condition which does
not now exist anywhere in the harbor. Flushing should have been better at that
time, before Pier J was built, which also indicates the degree of pollution prior
to environmental legislation.

Benthic Biomass

Biomass of benthic organisms can also be used to indicate environmental quality,
although the weight of a few large individuals in a sample can create a misleading impression. Biomass measurements were made in 1975 and 1976 in a study of a
proposed LNG channel (Fig. 9). This indicated that within approximately the 18-
foot contour closest to the waste outfalls and the shore, there was a zone of relatively low productivity or biomass (grams per square meter). In the outer harbor area studied, a zone of relatively high productivity was delineated (Fig. 10).
This study confirmed the distribution of species and numbers found in the earlier

TABLE I
DOMINANT ORGANISMS REPORTED FROM
LONG BEACH HARBOR IN AN INNER TO OUTER HARBOR ARRAY

	Very Polluted	Polluted	Semi-healthy II	Semi-healthy I	Healthy
Reish, 1959 ⟶	*Capitella capitata*	*Cirriformia luxuriosa*		*Polydora paucibranchiata* *Schistomeringos longicornis*	*Tharyx ? parvus*
Hill, 1974		"Polluted" (station 24)	*Capitella capitata* *Capitita ambiseta* *Polydora ligni*	"Healthy" (station 27)	*Tharyx parvus* *Cossura candida* *Haploscoloplos elongatus*
AHF, 1975		Group Z *Capitella capitata* *Armandia bioculata* *Polydora ligni* *Pseudopolydora paucibranchiata*	Group Y *Schistomeringos long.* *Capitella capitata* *Ophiodromus pugetten.* *Theora lubrica*	Group X *Euchone limnicola* *Callianassa* *Cryptomya calif.* *Nephtys c. franc.*	Group W *Tharyx ? parvus* *Cossura candida* *Haploscoloplos elongatus* *Prionospio pinn.* — Group V *Macoma acolasta* *Notomastus tenuis* *Prionospio pyg.* *Tellina modesta*
MBC, 1975			Group I *Tharyx* *Cossura candida* *Capitellidae** *Euphilomedes*	Group II *Tharyx* *Cossura candida* *Paraonis g. oculata*	Group III *Tharyx* *Cossura candida* *Paraonis g. oculata* — Group IV *Tharyx* *Euphilomedes carcharodonta* *Capitellidae** *Euphilomedes*

* *Capitella ambiseta* and *Mediomastus*.

Source: Port of Long Beach Master Environmental Setting (HEP, 1976).

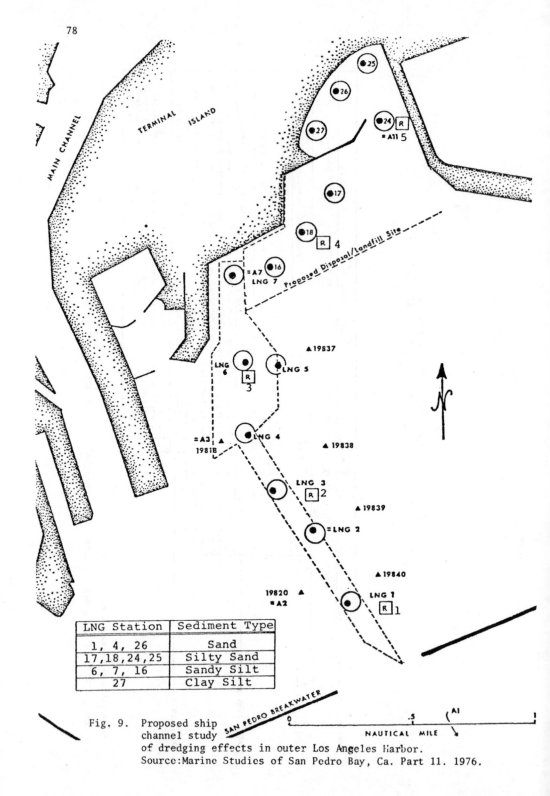

LNG Station	Sediment Type
1, 4, 26	Sand
17,18,24,25	Silty Sand
6, 7, 16	Sandy Silt
27	Clay Silt

Fig. 9. Proposed ship channel study of dredging effects in outer Los Angeles Harbor.
Source:Marine Studies of San Pedro Bay, Ca. Part 11. 1976.

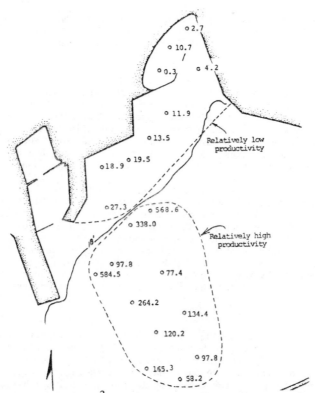

o 2.7

o 10.7

o 0.3   o 4.2

o 11.9

o 13.5   Relatively low
              productivity

o18.9 o 19.5

o27.3   o 568.6
    o 338.0
              Relatively high
              productivity

o 97.8
o 584.5   o 77.4

o 264.2
        o134.4

o 120.2

        o 97.8
o 165.3
    o 58.2

Fig. 10. Animal Biomass in g/m^2 at benthic Stations.   Source: Marine Studies of
San Pedro Bay, California.   Part 11.   1976.

study (Fig. 6) for part of the area of the waste effluent field.   The shallow area
not only contains the outfalls but wave action creates stirring of the unconsolidated sediments.   Dumping has occurred there extensively in the past as well.   In
a series of field experiments (HEP, 1976) racks of jars containing newly exposed
dredged surface sediments were placed at five locations (Figure 9, R symbols).   Although many of the jars were destroyed, presumably by ships' anchors and storms,
there was evidence of good potential for recolonization.   In the winter-spring periods of exposure, the best biomass gains were at locations just east of the TITP
outfall and within the zone of influence of the cannery wastes, where waters warmer.
In the summer, the shallow water stations showed very much higher biomasses than
the stations nearer the breakwater.   This contrasted considerably with the biomasses
and benthic samples taken from the sediments directly by grab sampler or boxcorer.
The recolonization jars are deployed in racks on the bottom sediments by divers,
and the difference between them and the grab samples may lie in the shelter provided
above the sediment by the jars.   Taken alone, the recolonization data were too few
to quantify the distribution patterns accurately, but they provide evidence that
effluent receiving waters and sediments are not toxic, but are really quite productive.

Phytoplankton Productivity

Phytoplankton productivity, photosynthetic pigments and assimilation ratio in the
general area of the outfalls have been monitored monthly since 1971.   These measurements offer and indication of the fertility of the associated waters in terms of

how much organic material is being produced by photosynthesis and how large a population of phytoplankton is involved in the production. The highest annual mean productivity and chlorophyll for 1973 and 1974 were found at the mouth of the Los Angeles River, east of Pier J in Long Beach.

The area around the waste outfalls showed moderate levels of productivity and chlorophyll with somewhat higher values than for the stations outside the harbor. Assimilation ratios, measurements of productivity per unit of population, showed patterns reflecting the enrichment associated with the effluents. Higher assimilation ratios were found in and around the outfalls area than in many other areas, though they were not the highest found in the harbor. The inner channels, to the northwest in Los Angeles and to the northeast in Long Beach, and the mouth of the San Gabriel River east of the harbors had higher mean annual values.

Seasonal variations in these measurements (Oguri, 1974) were similar in pattern to those occurring offshore, but varied in magnitude. A spring bloom, primarily of diatoms, showed higher values of productivity, pigments and assimilation ratio. In the summer and fall, secondary peak blooms occurred, usually of dinoflagellates. These were followed by the minimum values found in winter. The secondary blooms in 1971 through 1974 were notably more intense and longer lasting than those which occurred in 1975 and 1976. This may be due to the institution of more rigid waste discharge requirements by the Regional Water Quality Control Board or to decreased rainfall and milder temperatures than prevailed in the earlier years.

Zooplankton

Zooplankton in the Los Angeles-Long Beach Harbor has been reported on, following the two-year survey of the entire harbor (AHF, 1976). The zooplankton of the harbor consisted of virtually two distinct populations. The inner harbor had generally lower populations than in the outer harbor. However, the copepod *Acartia tonsa* comprises 69% to 80% of the total population in the inner harbor and also is prominent in the outer harbor. *Oithona oculata*, a cyclopoid copepod, also is found in abundance in the inner harbor but is much less apparent in the outer harbor. The cladocerans *Podon polyphemoides* and *Evadne nordmanni* are restricted almost exclusively to stations in the outer harbor. Figures 11 to 14 show computer maps of population densities based on settling volume for 1973 and 1974. The highest population densities for both years occurred in the area near the outfall, and may reflect the enrichment provided by the effluents.

Fish

More than 130 species of fish have been reported from Los Angeles-Long Beach Harbors (Chamberlain, 1973, 1974; Stephen, Terry, Subber and Allen. 1974; AHF, 1976) collected by trawl, gill net and hook and line (Figure 15). Stephens, et al. (1974), reported that the harbor supported a fish fauna richer than offshore areas of similar bottom and depth. The most numerous fish in the harbor were the white croaker and the northern anchovy, which made up 69% of the catch. Both species are plankton feeders and their large numbers were probably related in good measure to enrichment of the harbor by the effluents supplied.

Stephen, et al. (1974) described the distribution in the harbor as consisting principally of three general populations of fish based on otter trawl collections shown in Figure 16. Croakers 1) were distributed throughout the outer harbor, except where 2) the flat fish were found closest to the breakwater in the western areas of the outer harbor and 3) where rockfish abounded near the breakwater in the outer Long Beach Harbor. The croaker population appeared to be the most tolerant of the effluents discharged in the harbor. Fish occur in other areas of the harbor but cannot be sampled effectively by trawling. Since 1974, fish populations

ABSOLUTE VALUE RANGE APPLYING TO EACH LEVEL
(°MAXIMUM° INCLUDED IN HIGHEST LEVEL ONLY)

MINIMUM	0.55	1.31	2.06	2.82	3.57	4.33	5.08	5.84	6.60	7.35
MAXIMUM	1.31	2.06	2.82	3.57	4.33	5.08	5.84	6.60	7.35	8.11

PERCENTAGE OF TOTAL ABSOLUTE VALUE RANGE APPLYING TO EACH LEVEL

10.00	10.00	10.00	10.00	10.00	10.00	10.00	10.00	10.00	10.00

FREQUENCY DISTRIBUTION OF DATA POINT VALUES IN EACH LEVEL

LEVEL 1 2 3 4 5 6 7 8 9 10

Fig. 11-a. Mean plankton settling density, 1973.

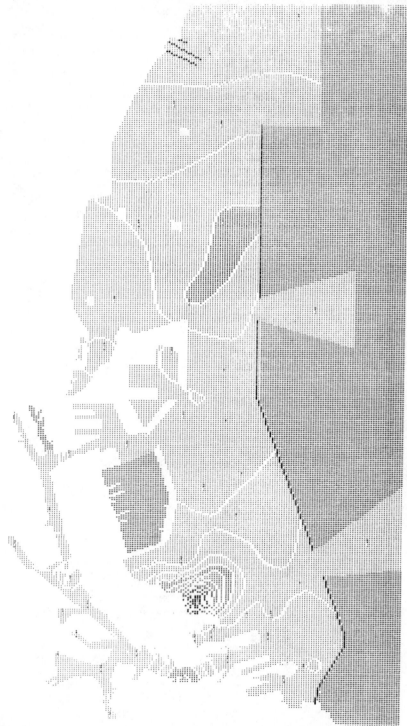

Mean Plankton Settling Density – 1973

Fig. 11-b. Mean plankton settling density, 1973.

```
ABSOLUTE VALUE RANGE APPLYING TO EACH LEVEL
     ('MAXIMUM' INCLUDED IN HIGHEST LEVEL ONLY)

  MINIMUM    0.42    0.67    0.92    1.17    1.43    1.68    1.93    2.18    2.44    2.69
  MAXIMUM    0.67    0.92    1.17    1.43    1.68    1.93    2.18    2.44    2.69    2.94

PERCENTAGE OF TOTAL ABSOLUTE VALUE RANGE APPLYING TO EACH LEVEL

            10.00   10.00   10.00   10.00   10.00   10.00   10.00   10.00   10.00   10.00

FREQUENCY DISTRIBUTION OF DATA POINT VALUES IN EACH LEVEL
  LEVEL        1          2          3          4          5          6          7          8          9          10
           ======================================================================================================
           .........  '''''''''  ---------  =========  +++++++++  XXXXXXXXX  OOOOOOOOO  @@@@@@@@@  @@@@@@@@@  #########
           .........  '''''''''  ---------  =========  +++++++++  XXXXXXXXX  OOOOOOOOO  @@@@@@@@@  @@@@@@@@@  #########
  SYMBOLS  ....1....  ''''2''''  ----3----  ====4====  ++++5++++  XXXX6XXXX  OOOO7OOOO  @@@@8@@@@  @@@@9@@@@  ####*####
           .........  '''''''''  ---------  =========  +++++++++  XXXXXXXXX  OOOOOOOOO  @@@@@@@@@  @@@@@@@@@  #########
           .........  '''''''''  ---------  =========  +++++++++  XXXXXXXXX  OOOOOOOOO  @@@@@@@@@  @@@@@@@@@  #########
           ======================================================================================================
  FREQ.        6          4          9          6          5          7          2          3          0          1
        1  I..1..I    I''2''I    I--3--I    I==4==I    I++5++I    IXX6XXI    IOO7OOI    I@@8@@I               I##*##I
        2  I..1..I    I''2''I    I--3--I    I==4==I    I++5++I    IXX6XXI    IOO7OOI    I@@8@@I
        3  I..1..I    I''2''I    I--3--I    I==4==I    I++5++I    IXX6XXI               I@@8@@I
        4  I..1..I    I''2''I    I--3--I    I==4==I    I++5++I    IXX6XXI
        5  I..1..I               I--3--I    I==4==I    I++5++I    IXX6XXI
        6  I..1..I               I--3--I    I==4==I               IXX6XXI
        7                        I--3--I                          IXX6XXI
        8                        I--3--I
        9                        I--3--I
```

Fig. 12-a. Mean plankton settling density, 1974.

84

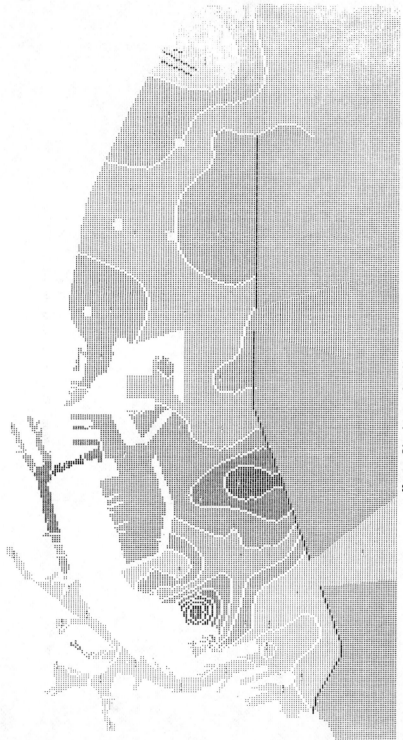

Mean Plankton Settling Density – 1974

Fig. 12-b. Mean plankton settling density, 1974.

```
ABSOLUTE VALUE RANGE APPLYING TO EACH LEVEL
      ('MAXIMUM' INCLUDED IN HIGHEST LEVEL ONLY)

  MINIMUM       0.0       1.13       2.26       3.38       4.51       5.64       6.77       7.90       9.02      10.15
  MAXIMUM       1.13      2.26       3.38       4.51       5.64       6.77       7.90       9.02      10.15      11.28


PERCENTAGE OF TOTAL ABSOLUTE VALUE RANGE APPLYING TO EACH LEVEL

               10.00     10.00      10.00      10.00      10.00      10.00      10.00      10.00      10.00      10.00


FREQUENCY DISTRIBUTION OF DATA POINT VALUES IN EACH LEVEL
  LEVEL          1         2          3          4          5          6          7          8          9          10
          ==================================================================================================================
           .........  '''''''''  ---------  =========  +++++++++  XXXXXXXXX  000000000  888888888  888888888  ■■■■■■■■■
           .........  '''''''''  ---------  =========  +++++++++  XXXXXXXXX  000000000  888888888  888888888  ■■■■■■■■■
  SYMBOLS  ....1....  ''''2''''  ----3----  ====4====  ++++5++++  XXXX6XXXX  0000700000 8888B8888  8888988888 ■■■■*■■■■
           .........  '''''''''  ---------  =========  +++++++++  XXXXXXXXX  000000000  888888888  888888888  ■■■■■■■■■
           .........  '''''''''  ---------  =========  +++++++++  XXXXXXXXX  000000000  888888888  888888888  ■■■■■■■■■
          ==================================================================================================================
  FREQ.          5         2          3          7          7          7          4          1          1          1
       1    I..1..I   I''2''I    I--3--I    I==4==I    I++5++I    IXX6XXI    I00700I    I88888I    I88988I    I■■*■■I
       2    I..1..I   I''2''I    I--3--I    I==4==I    I++5++I    IXX6XXI    I00700I
       3    I..1..I              I--3--I    I==4==I    I++5++I    IXX6XXI    I00700I
       4    I..1..I                         I==4==I    I++5++I    IXX6XXI    I00700I
       5    I..1..I                         I==4==I    I++5++I    IXX6XXI
       6                                    I==4==I    I++5++I    IXX6XXI
       7                                    I==4==I    I++5++I    IXX6XXI
```

Fig. 13-a. *Podon polyphemoides*, 1974.

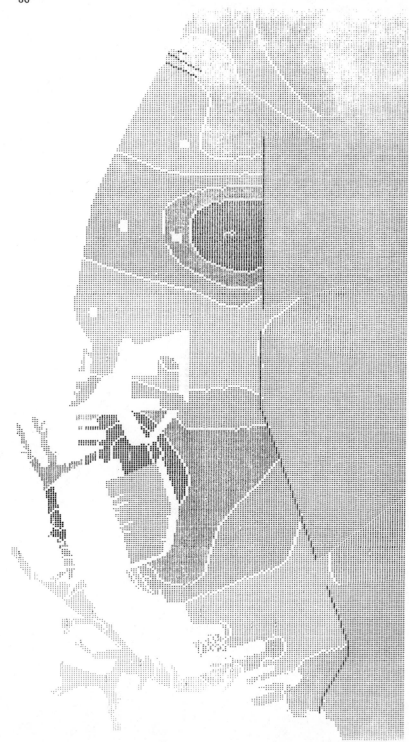

Podon polyphemoides — 1974

Fig. 13-b. *Podon polyphemoides*, 1974.

ABSOLUTE VALUE RANGE APPLYING TO EACH LEVEL
(*MAXIMUM* INCLUDED IN HIGHEST LEVEL ONLY)

MINIMUM	5.60	6.98	8.36	9.73	11.11	12.49	13.86	15.24	16.61	17.99
MAXIMUM	6.98	8.36	9.73	11.11	12.49	13.86	15.24	16.61	17.99	19.37

PERCENTAGE OF TOTAL ABSOLUTE VALUE RANGE APPLYING TO EACH LEVEL

10.00	10.00	10.00	10.00	10.00	10.00	10.00	10.00	10.00	10.00

FREQUENCY DISTRIBUTION OF DATA POINT VALUES IN EACH LEVEL

LEVEL 1 2 3 4 5 6 7 8 9 10

SYMBOLS . : | = + X O 8 ⊠ ■

FREQ. 1 2 3 4 5 6 7 8 9 10

Fig. 14-a. *Acartia tonsa.*

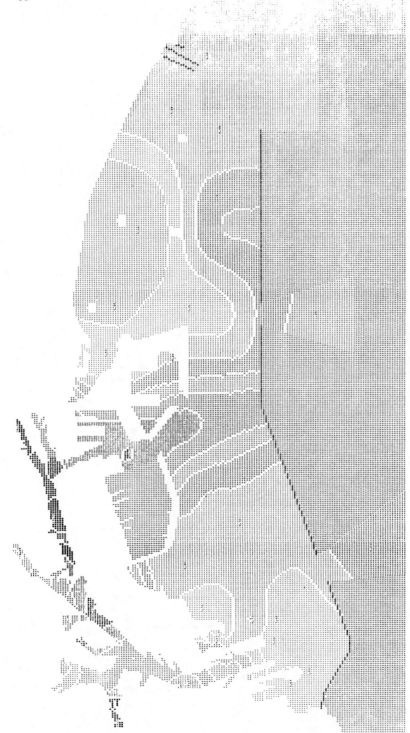

Acartia tonsa

Fig. 14-b. *Acartia tonsa.*

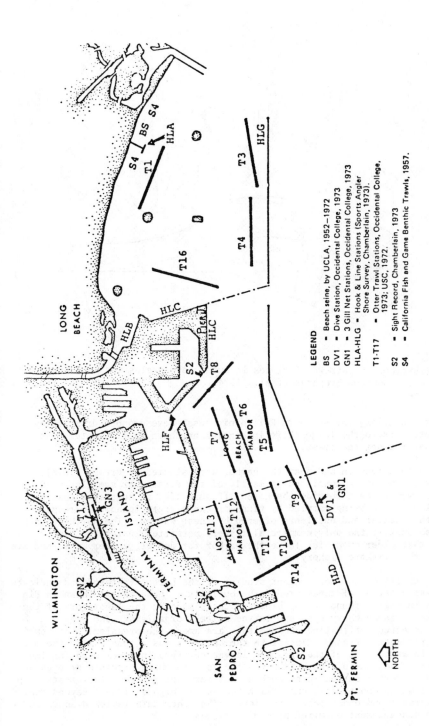

Fig. 15. Harbor fish surveys to 1974.

LEGEND

BS = Beach seine, by UCLA, 1952–1972
DV1 = Dive Station, Occidental College, 1973
GN1 = 3 Gill Net Stations, Occidental College, 1973
HLA-HLG = Hook & Line Stations (Sports Angler Shore Survey, Chamberlain, 1973).
T1-T17 = Otter Trawl Stations, Occidental College, 1973; USC, 1972.
S2 = Sight Record, Chamberlain, 1973
S4 = California Fish and Game Benthic Trawls, 1957.

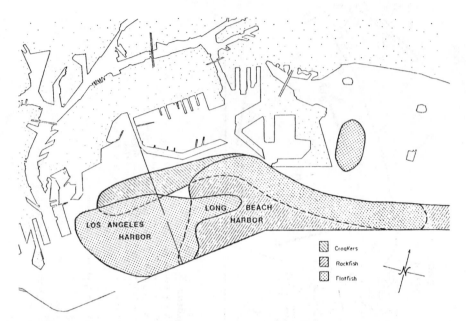

Figure 16. Distribution pattern of fishes in outer Los Angeles and Long Beach Harbors. Source: Stephens et al., 1974.

in the harbor and along the southern California coast have dropped drastically. Some attribute this to the large shift in central northern Pacific water masses and warmer water along the coast.

The anchovy populations in and near the harbor were studied by Brewer (1975). In the harbor the population consisted primarily of the first year age class. Since egg abundance was richest outside the harbor (Figure 17), it suggested that the juveniles either were being carried by tides or can migrate inshore, perhaps attracted by the shelter and nutrients offered by these waters. The northern anchovy caught as bait is the only commercial fishery permitted inside the harbor. It is surmised that older anchovies leave the harbor for deeper, colder water and are recruited to the offshore fishery.

Preliminary results of the more recent studies of the harbor ichthyofauna indicate that there was a series of zones around the outfalls. These zones, shown in Figure 18, were initially found during bioassay tests made in the laboratory of the cannery wastes using anchovy eggs and larvae as test organisms. The eggs and larvae were selected for testing because they are among the organisms most sensitive to environmental stress. Closest to the outfalls was a zone called the mortality zone, in which none of the eggs and larvae survived in water used for laboratory tests. The zone of inhibition was the zone in which exposure of eggs and larvae to the waters resulted in as much as 50% survival in laboratory tests. In the zone of lower productivity, survival noted was as high as 100%. Beyond that no mortality was found under test conditions. The other data reviewed above indicated that bioenhancement occurred beyond that zone.

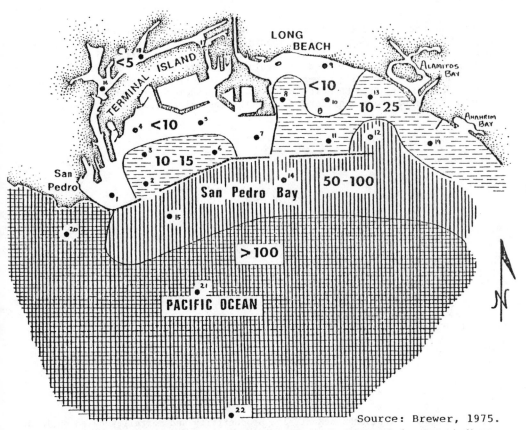

Figure 17. Map of the Los Angeles-Long Beach Harbors and San Pedro Bay, indicating the mean number of anchovy eggs per standardized trawl between Feb.
1973 and Sept. 1974.

The marine environment flourishes when the food web or ecosystem has many levels
of interaction. Uptake of protein, amino acid and glucose in the effluent waste
fields by bacteria has been documented (Sullivan, 1978). Fish gut contents were
examined and correlated with Stephens' earlier studies of fish distribution patterns. (McConaugha, 1976; Reish and Ware, 1976). Fish such as croaker and bottom feeders (flatfish) feed on the benthic polychaete worms, which in turn feed on
particulate organic detritus, flocculates and/or on bacteria filtered from the
water. Chamberlain (1975) and Bever and Dunn (1976) demonstrated that the amino
acids in cannery wastes can be assimilated directly by fish, which imbibe waters
and also pump them through their gills for aeration.

In unpublished studies on energy transfer as calories, E. Zerba and J.S. Stephens
determined that cannery effluents contain about 5500 calories per gram. Twenty-
five harbor organisms from four phyla were measured. The worms furnished the
highest mean energy values of the animal food available, at 4732 calories per gram.
This was followed in descending order by echinoderms, crustaceans and molluscx,
all of which require more energy expenditure by predators to capture and/or digest,
due to their mobility or their tougher exoskeletons. Small wonder that the fish
fed in large numbers on the worms in the effluent field. Even anchovies, which
are not normally bottom feeders, were found to do so on occasion in the harbor.

Research on modeling the capacity of the harbor to maintain adequate dissolved oxygen levels while accepting waste loading has been directed toward giving the canners and other processors of natural wastes such as sanitation districts, a tool that ultimately can be used to regulate nutrient (BOD) input according to the conditions of the receiving waters (Stanley-Miller, Kremer and Chian, 1976; Kremer, 1978). The Assimilation Capacity model is still under development; Figure 19 represents all the factors considered before the most significant ones were determined.

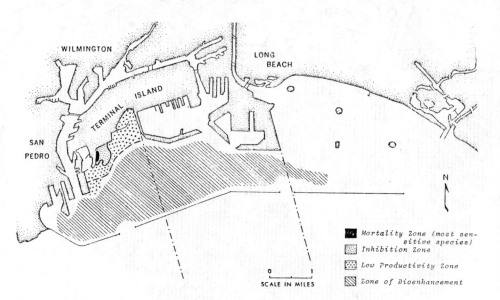

Figure 18. Zones of bioenhancement, low productivity and mortality based on laboratory bioassay tests. Source: Brewer, 1976.

Regulations

Federal laws are very general. It is the regulations adopted to implement the law that are enforced and it is the regulations that can be changed by Congress, in the light of new information. However, deciding what changes are desirable and getting legislation passed are two great difficulties. Enforcement practices have appeared to prefer one measurement, one rule, for the entire continental United States and all its islands, from Samoa and Hawaii to Puerto Rico; two areas as different as the Great Lakes and New Orleans come under the same rule. For enforcement purposes that may be fine; it is certainly convenient. For the environment and the economy, it may be very damaging.

The Environmental Protection Agency has interpreted its "one rule" mandate to include defining the "best available technology" as being "secondary waste treatment for all occasions." Point source control of non-nutrient toxicants is in fact far more important, if effluents can be managed such that the BOD loading does not impact the quality of the receiving waters. The capacity of each body of water receiving effluent can be determined and wastes managed accordingly. Furthermore, the policy does not allow for picking an ocean "solution" in a land-scarce area, nor does it really permit innovative developments.

While the industry and researchers were working toward a manage effluent, the Environmental Protection Agency overrode the California Water Quality Control Board and ordered the Terminal Island canneries to connect with the new TITP secondary waste facility. Heavy fines can be levied to force compliance, so the canners completed connection to TITP by January 1, 1978. The Los Angeles Regional Water Quality Control Board ordered a study for one year following hookup, to document any changes in the receiving waters during 1978; the investigations by HEP are in progress.

CONCLUSION

It must be recognized that EPA has the ultimate authority and only delegates powers to the California Water Resources Agency. They in turn developed policies including the Bays and Estuaries Policy. This states that 1) "persistent or cumalative toxic substances shall be removed ... to the maximum extent practicable through source control or treatment prior to discharge," and that 2) "... outfall or diffuser systems shall be designed to achieve the most rapid initial dilution practicable to minimize concentrations of substances not removed by source control of treatment..."

While the EPA has maintained that dilution is not the solution to pollution, it is a respected and viable method for managing wastes. Having been urged by Public Health to reduce intake of water, the canners were urged by HEP to use more sea water in their processing flow so that the valuable nutrient input to the harbor would still be available while the problem of inhibition at the outfall would be reduced or eliminated. Without the opportunity to make such modifications, the nutrient source was lost to the treatment plant microbial floc. This, then, becomes a problem of where to dump the sludge on landfill dumpsites.

Strangely enough, some Public Works staff involved with the Terminal Island Treatment Plant have agreed that a managed level of primary cannery waste effluent mixed with the new TITP secondary effluent would probably help to create and maintain a balanced ecosystem. It would incidentally have saved them numerous problems derived from trying to maintain a salinity-tolerant floc in the face of highly variable salinities in cannery wastes.

Perhaps the best evidence that bioenhancement can occur and that the nutrients of wastes are available, can be found in EPA's own news release of May 17, 1977, headed "EPA SAYS SOME POLLUTANTS USEFUL AS FOOD FOR UNDERWATER FARMS." According to the Federal Register, permits will be given for experimental projects for "underwater farms." Pollutants which might be used include ... those which are rich in nutrients or food, such as food processing wastes ... (or) some treated sewage plant discharges..."

"Research ... indicates that certain of these pollutants could be used ... for plant and animal species, including clams, oysters and shrimps, ... catfish, trout, pompano, salmon ... wild rice and aquatic plants..."

In short, if you fence off the receiving waters and raise plants and animals, effluents can be important nutrients. But the same system apparently is not valued in a harbor, where the organisms do the selection of their own habitat rather than a specialist representing human values who selects species for cultivation. Inquiries to obtain a permit for in-the-harbor mariculture were met with enthusiasm by some agency biologists but with flat rejection by EPA on the grounds that the canners were looking for a ruse to avoid hook-up.

94

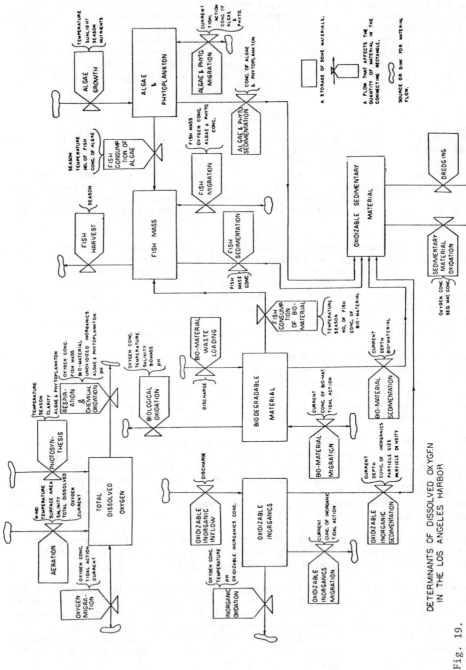

Fig. 19.
Factors considered in developing a model of the assimilation capacity (dissolved oxygen budget) of the receiving waters.

DETERMINANTS OF DISSOLVED OXYGEN
IN THE LOS ANGELES HARBOR

The best practicable treatment and the best available treatment for marine effluents have been defined as secondary waste treatment, at enormous future costs,
despite growing scientific evidence to the contrary from around the country. We
have yet to realize the burden of debt and taxation on future generations already
incurred by the massive federal loan program to fund construction. In cities
where there is no other recourse for potable water   is a necessary choice, but
along the oceans better use could be made of the wastes. EPA discovered that
traditional fresh water criteria based on BOD did not insure good water quality,
so receiving water quality was abandoned as the basis of compliance, and fixed
numbers for effluents were substituted. The biological quality of the receiving
waters should be the paramount criterion for marine discharges, rather than BOD or
other numbers.

The new EPA dredging regulation (EPA, 1977) reflect this and are based on bioassay
evaluation of the effects of dredged materials, a step in the right direction.
Even if it is more difficult to consider several modes of determining water quality and waste processing, it is essential that alternative strategies be considered
to insure use of limited and valuable nutrients.

LITERATURE CITED

Allan Hancock Foundation. 1965. Oceanographic and biological survey of the south-
    ern California mainland shelf. State Water Quality Control BD., Publ. 27,
    234 p; Appendix 446 p.
_____. 1976. Environmental investigations and analyses for Los Angeles-
    Long Beach Harbors. Final report to the U.S. Army Corps fo Engineers. 737 p.
Bascom, W. 1974. The disposal of waste in the ocean. Scientific American, 231(2):
    16-25.
_____. 1976. Waste solids entering the ocean in the Los Angeles area.
    Conference on Sludge Disposal Alternatives in the Ocean of Southern Californ-
    ia. Sept. 8, 1976. Calif. Instit. of Technology.
Bever, K., and A. Dunn. 1976. The energetic role of amino acid and protein meta-
    bolism in the kelp bass. In Marine Studies of San Pedro Bay, California.
    Part 12. Allan Hancock Found. and Sea Grant Prog., Instit. Mar. and Coastal
    Studies Univ. Southern Calif. p. 129-144.
Brewer, G.D. 1975. The biology and fishery of the northern anchovy in San Pedro
    Bay. Potential impact of proposed dredging and landfill. In Marine Studies
    of San Pedro Bay, California. Part 8. Allan Hancock Found. and Sea Grant
    Prog., Univ. Southern Calif. p. 23-44.
Chamberlain, D.W. 1973. Results of fourteen benthic trawls conducted in outer Los
    Angeles-Long Beach Harbors, California, May 24, 1972. In Marine Studies of
    San Pedro Bay, California. Part 2. Allan Hancock Found. and Sea Grant Prog.,
    Univ. Southern Calif. p. 107-145.
_____. 1974. A checklist of fishes from Los Angeles-Long Beach Harbors. In
    Marine Studies of San Pedro Bay, California. Part 4. Allan Hancock Found.
    and Sea Grant Prog., Univ. Southern Calif. p. 43-78.
_____. 1975. The role of fish cannery waste in the ecosystem. In Marine
    Studies of San Pedro Bay, California. Part 8. Allan Hancock Found. and Sea
    Grant Prog., Univ. Southern Calif. p. 1-22.
Chen, K.Y. and J.C.S. Lu. 1974. Sediment compositions in Los Angeles-Long Beach
    Harbors and San Pedro Basin. Marine Studies of San Pedro Bay, California.
    Part 7. Allan Hancock Found. and Sea Grant Prog., Univ. Southern Calif. 177 p.
Emerson, R.R. 1974. Preliminary investigations of the effects of resuspended sed-
    iment on two species of benthic polychaetes from Los Angeles Harbor. In
    Marine Studies of San Pedro Bay, California. Part 3. Allan Hancock Found.
    and Sea Grant Prog., Univ. Southern Calif. p. 97-110.
Goldman, J.C. and J.H. Ryther. 1975. Nutrient transformations in mass cultures of

marine algae. Environmental Egn. Div. Amer. Soc. Civil Eng., 101(EE3):351-364.

Harbors Environmental Projects. 1976. Potential effects of dredging on the biota of outer Los Angeles Harbor: toxicity, bioassay, and recolonization study. Marine Studies of San Pedro Bay, California. Part 11. 328 p.

Howell, J.A. 1977. Treatment of effluents. Oceans. May. p. 63-69.

Kremer, P.M. 1978 (in press). Dynamic oxygen model of Los Angeles Harbor receiving waters. In Marine Studies of San Pedro Bay, Califronia, Part 14. Allan Hancock Found. and Sea Grant Prog., Univ. Southern Calif.

McAnally, W.H. 1975. Los Angeles and Long Beach Harbors model study. Report 5. Tidal verification and base circulation tests. U.S. Army Corps of Engineers Waterways Experiment Station, Vicksburg, Miss.

McConaugha, J.R. 1976. Bioassay investigations of the impact of wastes on the copepod *Acartia tonsa*. In Marine Studies of San Pedro Bay, California. Part 12. Allan Hancock Found. and Sea Grant Prog., Instit. Mar. and Coastal Studies, Univ. Southern Calif., p. 215-225.

Oguri, M. 1974. Primary productivity in the outer Los Angeles Harbor. In Marine Hancock Found. and Sea Grant Prog., Univ. Southern Calif., p. 79-88.

Reish, D.J. 1959. An ecological study of pollution in Los Angeles-Long Beach Harbors, California. Allan Hancock Found. Occas. Paper 22. Univ. Southern Calif. 119 p.

_____. 1971. Effect of pollution abatement in Los Angeles Harbor, California. Mar. Pollution Bull. 2(5):71-74.

Reish, D.J. and R. Ware. 1976. The food habits of San Pedro Bay, California. Part 12. Allan Hancock Found. and Sea Grant Prog., Univ. Southern Calif. p. 113-128.

Robinson, K. and H. Porath. 1974. Current measurements in the outer Los Angeles Harbor. Marine Studies of San Pedro Bay, California. Part 6. Allan Hancock Found. and Sea Grant Prog., Univ. Southern California. 19 p.

Soule, D.F. and M. Oguri. 1972. Circulation patterns in Los Angeles-Long Beach Harbor. Marine Studies of San Pedro Bay, California. Part 1. Allan Hancock Found. and Sea Grant Prog., Univ. Southern Calif. 113 p.

_____. 1974. Data Report. Temperature, salinity, oxygen and pH in outer Los Angeles Harbor. Marine Studies of San Pedro Bay, California. Part 5. Allan Hancock Found. and Sea Grant Prog., Univ. Southern Calif. 76 p.

_____. 1976a. Physical water quality in the Long Beach Harbor area. Marine Studies of San Pedro Bay, California. Part 10. Allan Hancock Found. and Sea Grant Prog., Univ. Southern Calif. 173 p.

_____. 1976b. Bioenhancement studies of the receiving waters in outer Los Angeles Harbor. Mairne Studies of San Pedro Bay, California. Part 12. Allan Hancock Found. and Sea Grant Prog., Univ. Southern Calif. p. 3-58.

Southern California Coastal Water Research Project (SCCWRP). 1974. Annual Reprot.

_____. 1975. *Op. cit.*

_____. 1976. *Op. cit.*

_____. 1977. *Op. cit.*

Stephens, J.S., Jr., C. Terry, S. Subber and M.J. Allen. 1974. Abundance, distribution, seasonality and productivity of fish populations in Los Angeles Harbor, 1972-1973. In Marine Studies of San Pedro Bay, California. Part 4 . Allan Hancock Found. and Sea Grnat Prog., Univ. Southern Calif. p. 1-42.

Stanley-Miller, J., P.M. Kremer and W. Chiang. 1976. Quantification of the assimilative capacity of outer Los Angeles Harbor; a dynamic oxygen model. In Marine Studies of San Pedro Bay, Califonria. Part 12. Allan Hancock Found. and Sea Grant Prog., Univ. Southern Calif. p. 59-112.

Sullivan, C.W., A. Palmisano, S. McGrath, D. Kempin and G. Taylor. 1978 (in press) Microbial standing stocks and metabolic activities of microheterotrophs in southern California coastal waters. In Marine Studies of San Pedro Bay, California. Part 14. Allan Hancock Found. and Sea Grant Prog., Univ. Southern Calif.

Woodwell, G.M. 1977.   Recycling sewage through plant communities.   American Scientist.   65:556-562.

8

A COMPARISON OF THE ALCAN PLAN WITH THE
METHANOL APPROACH FOR PRUDHOE BAY NATURAL GAS

by

S. S. Marsden and M. T. Johnson
Stanford University

Utilization of Prudhoe Bay natural gas occupies the attention of both industry and government in the U.S. and Canada. The Alcan plan to build a 4800-mile long, 48-in diameter pipeline was endorsed in principle last year first by the Canadian government and then by that of the U.S. For political and environmental reason, it was chosen over the Arctic Gas and El Paso plans, both of which were more thoroughly designed and evaluated transportation systems. When the true costs are eventually recognized, however, this hasty political decision may be one which we live to regret simply because we cannot afford it.

There is considerable doubt as to whether or not the Alcan plan is the best way to bring Prudhoe Bay natural gas to market. It certainly has not undergone the extended engineering, environmental, and economic studies that the other two plans did. The published costs for it are climbing rapidly with time as the costs did for the Alyeska*oil pipeline and the Arctic Gas and El Paso plans. There is every reason to believe that its eventual cost would not be too different from the Arctic Gas plan because the total distances and the terrain crossed by both are quite similar. Since roads and rights-of-way are already available to some extent for the Alcan plan, the construction time should be shorter than for the Alcan plan.

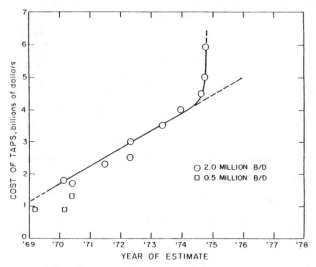

Fig. 1. Cost estimates for oil pipeline
at different times

But the lack of time to prepare the detailed engineering, environmental, and economic studies may delay construction greatly.

Several years ago we had the occasion to study the gradually increasing cost estimates for the Alyeska oil pipeline from the original one of $900 million in 1969 to the final one of about $8-9 billion for last summer (1). A linear graph of the data is given in Fig. 1 and a semilog graph in Fig. 2.

When the line for the latter graph was extrapolated from the time the study was completed to the time the pipeline was finished (summer of 1977), the cost predicted was very close to that currently reported.

More recently we have used the same approach for the three gas pipeline proposals (2). The relatively extensive data for the Arctic Gas plan is presented in the semi-log graph of Fig. 3, along with the limited and incomplete data for the Alcan plan. The even more limited data for the El Paso plan is given in Fig. 4.

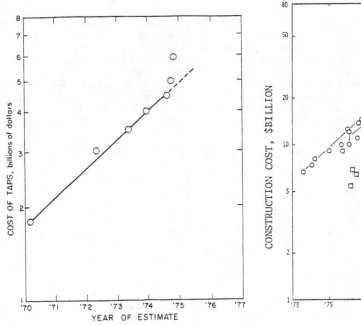

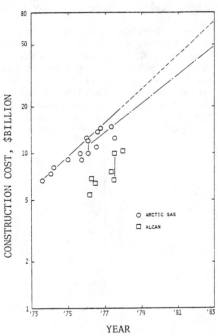

Fig. 2.   Cost estimates for oil
pipeline at different times

Fig. 3.   Announced costs for
Arctic Gas and Alcan
plans

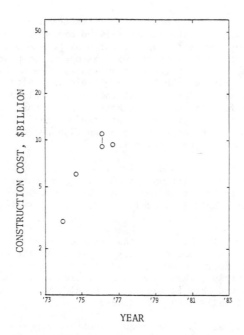

Fig. 4. Announced costs for El Paso Plan at different times

For the Arctic Gas plan the higher points at any given time represent costs for
the completed transportation system, while the lower ones are probably for only
portions of it. A line has been drawn through these higher points and extrapolated
into the future. In some places in the literature the construction time is given
as 1983. If this line is extrapolated to the latter date, an incredible cost of
$70 billion is obtained! Even if a line is drawn through the bulk of the points
and extrapolated to 1983, a figure of $50 billion is obtained. Considering that
the gas line will be six times as long as the oil line, it is not surprising that
its cost would be at least six times the $9 billion for the oil line.

It is apparent that there is not enough data to draw comparable lines for either
the El Paso or the Alcan plans. However, the length of the pipelines proposed for
both the Arctic Gas and the Alcan plans are about the same, and the terrain crossed
by both is similar. Thus, to a certain extent, the calculated cost data for the
Arctic Gas plan can be used to estimate the probable cost for the Alcan plan. This
strongly suggests that the cost at completion for the Alcan plan will be several
times greater than the figures now presented in the literature.

The purpose of this paper, however, is not to compare and contrast the three plans
already mentioned, but instead to describe briefly a fourth proposal which we have
worked on for several years. We shall also compare the Alcan plan with our proposal from the standpoints of economics, environmental impact, estimated completion
time, and utilization of product.

Briefly, our proposal is to convert the natural gas into methanol at Prudhoe Bay
by using standard, well-known processes, and then add the methanol to the Alyeska
oil pipeline already operating from there to Valdez. Methanol is a clear, color-
less, combustible liquid that can be used in internal combustion engines, in making plastics and a host of other petrochemicals, in growing single-cell protein,
in peak-shaving for electric power generating plants , and in a variety of other

ways. The conversion plants could be readily constructed and put into operating
condition on concrete barges in the industrialized areas of the U.S. or Japan, and
then towed to Prudhoe Bay during the brief, north slope summer. The barges could
be beached or moored at the shoreline and used as operation pads without ever dis-
mantling the plants and reconstructing them on shore. The methanol could be transported in the unused capacity of the Alyeska oil pipeline either in batches or else
by dispersing the crude oil in the methanol to form an emulsion-like fluid. In
either event, it can be further carried to market in ordinary tankers just as
crude oil is now transported. Our studies at Stanford indicate that dispersions
of crude oil-in-methanol may be superior to batch flow due to their better pumping
and tanker on-and-off loading characteristics.

In 1975 an engineering, economic, and environmental study was carried out on the
methanol approach for a transportation system built from scratch (3). We have now
modified this for the existing situation in Alaska, namely, a completed oil line
and our present need for a better natural gas utilization system than the Alcan
plan.

For the methanol approach, it is obvious that another pipeline will not have to
be constructed. There is considerable excess capacity in the Alyeska oil line
now operating which will increase when the additional pumps for which it was designed are installed. Because this line is classified as a common carrier, it
should transport any liquid that it can safely handle, and there is every reason
to believe that methanol falls into this class.

Because the methanol approach does not require another pipeline in addition to the
Alyeska oil line, the major capital expense would be the methanol plants. van
Poollen (4) has estimated that "...the optimum rate of gas production is 2 billion
cubic feet per day," but more recently, Doscher (5) has indicated that this is
probably too high from the standpoint of maximum oil recovery. This is less than
the amount requested for the several gas pipeline proposals because they also count
on taking gas from the gas cap, which would ultimately decrease oil recovery. Boyd
(6) has summarized the results of seven recent studies on the capital cost of methanol plants of various sizes, and has found that these all come very close to being
proportional to $850 x 10^6$ for 10^6 MM BTU/D capacity. These are given in 1976
dollars; if we apply the inflation rate of 10% per year used by him and others,
then the cost for 1978 is $1029 x 10^6$ for the same capacity. If we make the usual
conversion of 1,000 cu ft = 1 MM BTU for natural gas and multiply these three figures together, then the cost of methanol manufacturing facilities is $2057 x 10^6$,
or slightly more than $2 billion in 1978 dollars.

According to Boyd, these estimates "...include storage, mooring and loading facilities" in various parts of the world. We believe that these would be comparable
to the barge costs for the methanol approach at Prudhoe Bay because the buoyancy
chambers in these barges can be used for storage of both methanol and process water used in the synthesis.

Unlike the Alcan plan, one of the advantages inherent in the methanol approach is
that its implementation would be incremental. To elaborate upon this, one must
consider both pipeline capacity and gas production rates. To maximize the recovery of oil from a reservoir such as Prudhoe Bay, the oil production rate must be
the dominant factor and the gas production rate varied appropriately (5). But,
to be most economical, a gas pipeline must operate at or near capacity. Thus the
gas supplied to a pipeline would change considerably with time. To transport the
generally increasing gas production rate with time and without flaring, excess
capacity must be built into the pipeline system at large initial expense. But
with the methanol approach, additional conversion plants can be installed as needed

over a period of time, and eventually removed to other locations in the world several decades later, when no longer needed. In other words, the methanol approach is incremental, while a gas pipeline -- except to the extent that additional compressors add more capacity -- is not. The savings in implementation time and capital should be obvious. Therefore, the $2 billion for methanol plants given earlier is the upper limit of an incremental cost.

The methanol plants also require large quantities of fresh water which will probably have to be obtained at least in part from desalinization of sea water. Because the amount of surface water available at Prudhoe Bay is unknown, we have not been able to assess the size and cost of plants to do this.

In 1975, Hooker (3) estimated an operating expense for the dispersion of crude oil-in-methanol of $1.26/Mcf. We have estimated an operating expense for batch flow of $1.00/Mcf. The former is somewhat higher due to special equipment and materials required to disperse oil in methanol. The most recent operating expense announced for the Alcan plan is $1.09/Mcf (7); however, since this has been increasing significantly with time, the expenses for both the methanol approach and Alcan plan are nearly equivalent within the accuracy of the estimates.

From the environmental standpoint, a comparison must be made between the installation and operation of methanol plants at Prudhoe Bay and the construction of an entirely new pipeline across large wilderness areas. We believe that the former will be less adverse. Apart from the transportation system the availability of large quantities of methanol would have a positive effect on environmental improvement. There is evidence that its use, either directly or in blends with gasolines, leads to lower air pollution than from gasoline or diesel fuel alone. It is being considered in California not only as a fuel for peak load shaving, but also for use when air pollution becomes particularly bad due to burning high sulfur fuel oil in electrical generating plants (6). The Alcan plan would provide a low air pollution fuel to a wide area, while the methanol approach would provide one to a somewhat different one. As is the case with all environmental matters, one must consider and try to balance all factors; we merely present some of those here that we believe are important and let the reader reach his or her decision.

The predicted completion time of the Alcan pipeline seems to be some uncertain time in the future. The availability of roads along the proposed route in Alaska, most of Alberta, Saskatchewan, and most of the lower 48 states should help, but there are still major portions that traverse wilderness areas. The production, transportation, and installation of 4,787 miles of what would be the longest and most expensive pipeline in the world would require massive mobilization of manpower and equipment.

Delivery time for the first methanol plants and for the concrete barges should be one to two years. Moving them to site, however, can only be done during a short period in the summer, which, of course, affects installation time.

In closing, we should like to respond to several points that have been raised about the methanol approach. Some have said that a large amount of the energy in the gas is lost during the conversion process to mentanol. This is true if chemically pure methanol is produced, but apparently not so if other natural gas components (ethane, propane, butane) are also converted to a mixture of alcohols. The methanol approach allows transportation of these valuable components of produced natural gas to market. It is doubtful if all of the butane and the propane can be transported in Alyeska's hot oil line or Alcan's cold gas line.

Some have said that the methanol approach would produce more of these petrochemicals than we now consume. This is true, so new uses such as those mentioned above and others will have to be developed. Closing down methanol plants in the lower 48 states which are now running on natural gas would free this gas for more urgent and local uses. In some cases, these plants might be rebuilt on barges and sent to Alaska. Mobil Oil has demonstrated that methanol can be converted to gasoline, presumably at acceptable costs (9). Methanol can also be converted into a pipeline gas again, but the energy cost of this is significant (9).

The methanol approach provides us with an important energy option. The west coast has a much sounder position in regard to domestically produced crude oil than does the east coast. Pipelines or tankers to get western oil east are both very expensive and are encountering serious environmental obstacles. We should reconsider a proposal made several years ago to develop submarine tankers which would transport not simply crude oil from Prudhoe Bay to the east coast, but dispersions of crude oil-in-methanol. Technically, this appears to be feasible, and economically it looks attractive (10).

We are not presenting the methanol approach as a plan that has been worked out and evaluated in every engineering, economic, and environmental detail, but rather as a very promising approach that should be carefully and thoroughly evaluated in the near future. We need to be ready with an alternative plan when the Alcan plan falters for one of many reasons.

REFERENCES

1. Hooker, P.R., and Marsden, S.S.: "What Will Oil and Gas Transportation from Alaska Eventually Cost?" The Stanford Earth Scientist, Win.-Spr. 1976, p. 3.

2. Johnson, M.T., and Marsden, S.S.: "The Future for Prudhoe Bay Natural Gas," unpublished manuscript.

3. Hooker, P.H.: "Transportation of Cold Crude Oil and Natural Gas from the Arctic as a Cold Dispersion of Oil in Methanol," Ph.D. dissertation, Petroleum Engineering Department, Stanford University (1975).

4. Anon.: "Alaska Natural Gas Transportation System, Final Environmental Impact Statement, Overview," Mar. 1976, U.S. Dept. of Interior, Washington, D.C. p.5.

5. Doscher, T.M.: Testimony before Energy & Natural Resources Committee, Congressional Record - Senate, Oct. 28, 1977, S18112.

6. Boyd, R.: "Methanol as a Transport Alternative for Remote Natural Gas," Cal. Energy Resources Cons. and Dev. Com., Professional Paper No.1, Jan. 11, 1977 (draft), p. 22.

7. The Oil and Gas J., (July 11, 1977), p. 56.

8. Anon.: "Mobil Proves Gasoline-from-Methanol Process," C&EN, (Jan. 30, 1978), p. 26.

9. Shirtum, R.P. et al.: "Conversion of Coal-Based Methanol to Gaseous Fuel Proposed," Oil & Gas J., (Oct. 31, 1977), p. 106.

10. Personal Communication, W.H. Kumm, Arctic Enterprises, Severna Park, Maryland.

* Legal name of pipeline company

SECTION III

ENVIRONMENTAL CONSTRAINTS AND POLLUTANT ANALYSIS

Variability in Response of Marine Organisms to Petroleum
under Different Conditions

The Use of GC/MS in the Identification and Analysis of Organic
Pollutants in Water

Development of a Novel Hydrocarbon-in-Water Monitor

Has the Clean Air Act Really Stopped Industrial Growth?

9

VARIABILITY IN RESPONSE OF MARINE ORGANISMS TO
PETROLEUM UNDER DIFFERENT CONDITIONS

by

Dale Straughan
Institute for Marine & Coastal Studies

The variability in the response of different organisms to exposure to petroleum
under different conditions is important to:
A) interpret the results of standard toxicity tests
B) predict the impact of oil spills and design response plans
C) design and establish baseline parameters against which to measure impacts of
    exposure to petroleum.
It is acknowledged that it is impossible to test every species and every combination of variables that may influence its response. However, under the present
system of heavy reliance on standard toxicity testing data with little or no al-
lowance for variations in the field, it is possible that useful substances are
being banned and harmful substances authorized by regulatory agencies. It is also
possible that the wrong parameters could be measured as baselines and in the determination of the impact of oil spills. The purpose of this paper is to describe
the type of variability that should be considered in the operations listed above.

Straughan (1972) briefly discussed a number of factors which operate in field
conditions and which influence the impact of oil spills. These include: chemical
composition of the oil; dosage of the oil (volume related to contaminated area);
physical state of the oil; physical features of the oil spill site; weather conditions at the time of the spill; season of the year; previous exposure to petroleum; exposure to other pollutants; and cleanup methods employed. Since that time,
Kanter (1974) and Straughan (1976), have further demonstrated the variability in
response of the mussel, Mytilus californianus on a by-site basis, bimonthly basis
and annual basis. Straughan (1976) also emphasized the need to consider heterogenity in field populations. This heterogenity could be in terms of genetic
variability, reproductive variability, age, variability in tolerance, and/or variability in exposure to other stresses. Dicks (1976) has demonstrated the importance
of behavioral patterns in toxicity testing and ecological predictions.

I now wish to discuss this problem in terms of response that are not incorporated
in the standard testing procedures but which should be considered in data evaluation.

1. Escape or Avoidance Reactions of the Organisms

After the Santa Barbara oil spill, it was observed that the mussel, M. californianus
only opened its shell for periods during high tides in oil contaminated areas
(Brisby, pres. com.). The shell was closed during the periods when the mussels were
exposed to the highest concentrations of the floating oil. The animals, by this
response, were actually reducing their exposure to the minimal level. Therefore,
prediction of impact on these mussels on the basis of standard toxicity tests using
ocean surface oil concentrations would be very unreliable. Other avoidance mecha-

nisms include migration out of an area by mobile species.

Gastrapods such as Littorina spp. retreat into their shells and close the operculum on exposure to petroleum. They then fall from the substrate and will be lost from the population (Stein et al 1978).

2. Delayed Mortality

Most standard toxicity tests are based on a 96 hour exposure period with no follow-up period to record delayed mortality. Craddock (1977) in noting this limitation reported that Brungs (1973) recorded direct lethal effects several months after exposure to oil. One of the problems in determining mortality after an oil spill is when do you count the dead bodies? Soft bodied species and species not attached to the substrate decompose and/or are washed away rapidly. Other species such as bivalves and barnacles will remain on the substrate and can appear to be alive for some weeks but in reality the animals are dead and the valves are closed.

Let me demonstrate this problem further by reference to experimental data obtain in experiments with mussels, Mytilus edulis, collected from Fishermans Cove at Santa Catalina Island, Co.

Two groups each of 20 experimental animals, one in seawater and one in Santa Barbara crude oil from the Dos Cuadros field, were placed in a water bath at ambient sea-water temperature (23°C), gradually raised 10°C over the next hour, maintained at this temperature for 2 hours. The animals were then wiped clean and placed in a water bath with a flow through seawater system and mortality measured. Field measurements indicate, that in a tide pool situation where mussels were left str-anded in oil at low tide, the rise in temperature can be as rapid as this and can even go as high as 45°C over a 3 hour period before the return of the tide. Sea-water in contrast, does not heat as rapidly so that while the data does not reflect a field situation for a normal rock pool with seawater in the mussel beds, it is good approximation for conditions that can occur if the pool becomes filled with Santa Barbara crude oil at low tide on a summers day.

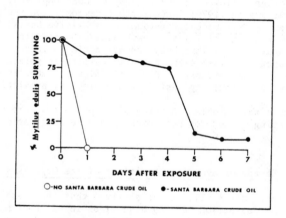

Fig. 1 Survival of the mussel, M. edulis after exposure to
elevated temperatures and petroleum.

Immediately after an experiment of this type it is frequently difficult to determine which mussels are living and which are dead. Therefore the mussels are
usually cleaned and placed in a water bath overnight and then examined. If the
data are considered 1 day after exposure (Fig. 1), the results would suggest an
80% survival among mussels in the oil in comparison to 100% mortality among mussels
in seawater. Even 4 days (96 hours) after exposure, there was still 75% survival
among the mussels exposed to petroleum. One day later, this was reduced to 15%
survival in M. edulis exposed to petroleum. In other words different conclusions
could be drawn depending on the time of termination of the experiment. If the experiment were terminated as soon as feasible after exposure (1 day) the data could
be interpreted to suggest that petroleum in some ways actually assisted survival
at these temperatures. A similar interpretation could also be made after 96 hours.
However a day later, the data show that there is probably no significant difference
between survival rates under either experimental condition.

Similarly in considering these data in relation to field conditions, data recorded
one to four days after exposure would indicate a high survival rate among mussels
in rock pool filled with oil at low tide, while data recorded on the fifth day
show that the mortality rate under these conditions is probably high.

3. Habitat Differences

Here, there are differences between the habitat of the animals in the field and
the laboratory as well as variations in habitat in the field to be considered.
Fig. 2 demonstrates the different patterns of exposure to petroleum at different
intertidal levels. For example, while animals at high tide might be exposed to
a stranded layer of petroleum for almost the entire tidal cycle, those at mid-tide
(level A) could be exposed to the stranded layer 50% of the time while those at
low tide, (level B) would barely be exposed to this stranded layer.

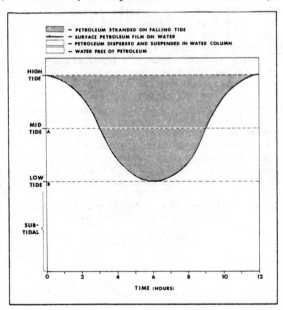

Fig. 2   Diagram to demonstrate different types of exposure to
petroleum at different intertidal levels.

Field evidence from a survey five months after the Metula oil spill in the Straits of Magellen provide several examples of differential impact of petroleum at different intertidal levels (Straughan, 1977). Field records on the distribution of the mussel Mytilus edulis chilensis at site G showed that the species was living at three low intertidal quadrats with no visible evidence of mortality at the time of the survey. A fourth quadrat extended up to the upper intertidal limit of distribution of this species and also bore an oil sheen on intersitial water as well as some evidence of petroleum still under the rocks. In this quadrat, very few living M. edulis chilensis were collected, while byssus threads with no shells attached were observed on the rocks. Such byssus threads without shells were not observed in three lower intertidal quadrats. This suggests recent mortality near the upper intertidal limit of distribution of this species but not in the low intertidal population. This could be due either to differences in tolerance at different intertidal levels and/or differences in exposure to petroleum at different levels.

The differences in exposure at different intertidal heights were also dramatically demonstrated during this survey. At site I for example a flat low intertidal area, over 1000 feet in width was covered with several inches of brown mousse. At the same site there was an area of asphalt like deposits in a higher intertidal area, while at high tide there were discontinuous deposits of wet black oil. Another example was the upper intertidal area in the marsh at Site C where there were two rows of stranded petroleum with visibly clean sediment between them. Chemical analyses of sediments from the two petroleum contaminated areas and the visibly clean sediments revealed no petroleum in the visibly clean sediments and different levels of petroleum contamination in the two strandings (5%, 25%).

4. Variability in Tolerances of Field Organisms

This can occur for a number of reasons including different stages in life and/or reproductive cycle, season of year, genetic variability, but also due to other external factors operating at different sites or at different times at the same sites. Several papers (Kanter, 1974, Straughan, 1976) have demonstrated that the mussel M. californianus, for example, appears to have a higher tolerance to low level exposure to Santa Barbara crude oil after the animals have been naturally chronically exposed to petroleum at Coal Oil Point than have M. claifornianus form other sites not chronically exposed to petroleum.

Since most harbors are subjected to at least intermittent pollution it is important to determine if there is a difference in the tolerance between organisms living within harbors and the same species not in harbors. To this end, a series of experiments were conducted in which the mussel Mytilus edulis collected inside and outside harbors were exposed to gradually elevated temperatures over a three hour period. Data in Table 1 show that M. edulis collected inside both Los Angeles and Santa Barbara Harbors were less tolerant to temperature changes than animals

Table 1. Mortality (%) in Mytilus edulis exposed to different water temperatures

	28-45°C	12-17°C
Los Angeles Habor		
Inside	100	0
Outside	20	0
Santa Barbara Harbor		
Inside	100	40
Outside	20	0

111

collected on the outside of the breakwater. Animals used in these experiments were 10 to 35 cm long and were monitored for mortality for five days.

In a second series of experiments larger M. edulis (40 to 50 cm in length) collected inside Los Angeles Harbor and from Fisherman Cove at Santa Catalina Island were gradually exposed to elevated water temperatures over a 3 hour period. The animals were subsequently monitored for mortality over a seven day period. Animals from Los Angeles Harbor were more susceptible to elevated, seawater temperature than those collected from the non-harbor environment at Santa Catalina Island (Table 2).

Table 2. Mortality (%) in M. edulis exposed to
different water temperatures

Temperature Range (°C)	Los Angeles Harbor	Santa Catalina Island
18-19	5	0
18-24	0	0
18-29	0	0
18-34	65	0
18-39	100	15

In a third series of experiments, two species of mussels, M. edulis and M. californianus were exposed to Santa Barbara Crude oil for different time periods and at different temperatures. The animals were then monitored for mortality over a five day period. M. edulis from Los Angeles Harbor was again less tolerant to stress at elevated water temperatures than M. edulis from the non-harbor environment (Table 3). However, the second species of mussel tested was even less tolerant of the stress provided by petroleum and temperature than M. edulis collected in either locality. M. californianus retained oil within the shell in some instances while M. edulis did not. This probably indicates a greater effective exposure to M. californianus than M. edulis due to such differences as better closure of shell and/or ability to live in a closed shell for longer periods.

Table 3. Mortality (%) in mussels exposed to Santa
Barbara crude oil

Exposure Temperature (°C)	Time (hrs)	M. edulis Collected Los Angeles Harbor	M. edulis Collected Santa Catalina Island	M. californianus Collected Santa Catalina Island
19-20°C	3	0	0	25
19-20°C	21	5	5	50
19-43	3	30	0	50
19-43	21	65	35	65

Experiments with Littorina scutulata collected from localities along the west coast of North America indicated that these animals were most susceptible to stress in terms of exposure to several types of oil and elevated temperatures of 29°C when collected from the extremes of their normal water temperature range (Fig. 3). Mortality was 100% in animals collected from sites when the seawater temperature was below 8°C and at 19°C, and mortality was lowest at 10°C (Straughan and Hadley, 1978).

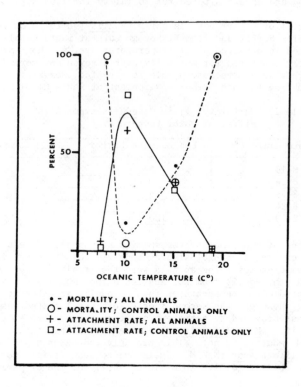

Fig. 3.   The relationship between mortality and attachment
rates of Littorina scutulata and field water temperatures at the time of pollution.

5.   Synergistic Effects

This can involve the sequential on concurrent exposure of an organism to more than
one stress and probably explains why M. edulis from Los Angeles Harbor are less
tolerant to exposure to elevated temperatures than M. edulis from Santa Catalina
Island.   Comparison of data obtained in experiments on M. edulis from Santa Catalina
Island involving elevation of temperature and exposure to Santa Barbara Crude oil
over a three hour period, indicated exposure to oil with no elevation in temperature
caused no mortality while elevation in temperature with no exposure to oil caused
15% mortality (Fig. 4).   However, when the animals were exposed to petroleum and
elevated temperature, there was 100% mortality.   In other words animals covered
with oil at low tide in the sun on a hot summer's day, would probably die while
those not exposed to oil and those exposed to oil but in the shade probably would
not die.

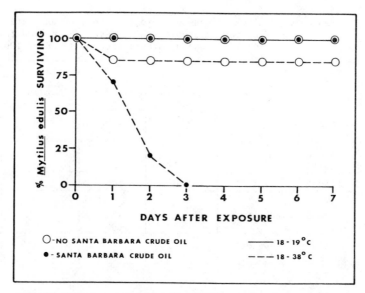

Fig. 4. Survival of M. edulis from Santa Catalina Island
         after exposure to elevated temperatures and petro-
         leum.

6. Sublethal Effects

These include such things as inability of animals to attach to the substrate, in-
terruption of breeding cycles, feeding mechanisms and escape mechanisms. In the
first instance for example, even though littorinid snails may be able to survive
exposure to a pollutant, if they cannot attach to the substrate they will be
washed away by wave action and the population lost from the site (Table 4). Note
that at 17°C, Littorina scabra, all survived exposure to Santa Barbara crude oil
either alone or mixed with the dispersant BP1100X but a maximum of only 3% and
12% respectively could have been considered to have remained attached to the substrate in the week following the experiment. Likewise at 29°C, a temperature
which is recorded in Naples, Florida where these animals were collected, while all
animals survived exposure to the dispersant, less than three quarters of them remained attached to the substrate. Under these experimental conditions while the
animals survived the pollutant they would probably still be lost from the population.

Another similar example is provided by changes in the ability of mussels to produce byssus threads (Fig. 5). In the example provided in which Mytilus californianus was exposed to Santa Barbara crude oil that had weathered for 8 days, for
a three hour period during which the temperature was gradually raised to 33°C, both
the control and experimental animals were initially under some stress after exposure because only 60% of the control and none of the experimental animals were
initially attached to the substrate by byssus threads. If the population were
further stressed by heavy wave action or surf during this period, probably the
population would lose more animals than are indicated by the final mortality figures
of 20% when exposed to elevated temperatures and 70% when exposed to this type of
oil and elevated temperatures.

Table 4. Minimum attachment rate (%) and survival rate (%)
of Littorina scabra over a seven day period following
exposure to Santa Barbara crude oil and BP1100X

	17°C		29°C	
	Minimum Attachment	Survival	Minimum Attachment	Survival
Control a	100	100	96	100
b	100	100	80	100
Santa Barbara Crude Oil	12	100	0	0
BP1100X	100	100	0	0
Santa Barbara Crude Oil: BP1100X 10: 1	3	100	0	0
BP1100X: Seawater 1: 10	96	100	72	100

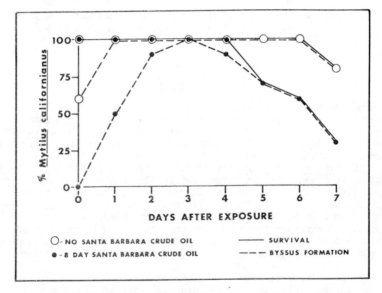

Fig. 5. Comparison of survival rates and byssus thread
formation in mussels, M. californianus after
exposure to elevated temperature (33°C) and
petroleum weathered for 8 days.

It has also been noted in our laboratory experiments that both species of Mytilus
will release gametes if exposed to Santa Barbara crude when they are in breeding
condition. This was also recorded in the field after the Santa Barbara oil spill
(Brisby, pres. com.). Exposure to Santa Barbara crude oil also stimulated the
release of gametes from the algae, Hysperophycus harveyanus, in laboratory experi-

ments. This would surely actually reduce the reproductive potential in species with this reaction, because the newly released gametes would probably be suscept-ible to exposure to petroleum.

7. State of Petroleum

While the chemical composition of the spilled petroleum is generally considered, frequently factors such as weathering and loss of some components and the state of the petroleum are ignored. For example, data presented in Figure 6 would suggest an increasing survival rate of M. edulis with increasing weathering of the petro-leum. Some indication of the loss is the fact that 10% by volume was lost from 1 gallon of this oil 1 over a 24 hour period at 20°C.

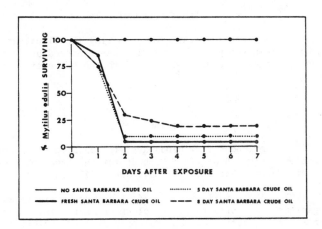

Figure 6. Comparison of survival rate of M. edulis when exposed to petroleum that had been weathered for different time periods.

Another factor is mousse formation. Most toxicity experiments using Kuwait crude oil, for example, are conducted using the crude oil mixed with seawater. However, Kuwait crude frequently forms a water in oil mousse which probably is a far more toxic formation than oil floating on or dispersed in water because it will simply remain covering many organisms. The large flat lower intertidal areas in the First Narrows of the Straits of Magellan remained covered in several inches of mousse, even at high tide, for many months (5+) after the Metula oil spill. No organisms were collected in quadrats in areas covered this manner, while 17 inver-tebrate species were collected from an adjacent quadrat which was not covered with mousse (Straughan, 1977).

In this discussion, I have demonstrated the variability in response of biota to exposure to petroleum and other related stresses. In some of these instances the response of the animal increases the survival rate under field conditions while in other instances survival of field populations will be reduced by the response of the organism. Genetic variability within a species itself is probably one of the most important mechanisms in survival of a species when exposed to any stress.

In areas such as Coal Oil Point, there could be selection among the settling larval and/or already settled organisms which allow for those individuals with best ability to survive chronic natural exposure to petroleum to have an advantage over those with the least ability to survive this stress. Hence, organisms from this area show a greater tolerance to laboratory exposure to petroleum than organisms collected from areas where this type of selection is not occuring.

Therefore in order to evaluate the data from standard toxicity tests for use in field situations, it is necessary to have basic biological and ecological data. These types of data are gradually being amassed through many avenues of research. However, it is still a source of wonder that baseline studies do not always include a segment to determine the type of tar associated with the biota and not just the volume of tar and that biotic data are being recorded without associated natural abiotic data.

ACKNOWLEDGEMENT

I wish to thank D. Hadley and T. Licari for preparing the illustrations, and P. Van Sciver for typing the manuscript. The research was financed by NOAA Office of Sea Grant, Department of Commerce Grant No. 2-35227 to the University of Southern California.

References

Brungs, W. A. 1973. Continuous flow bioassays with aquatic organisms: Procedures and applications. In: Biological Methods for the Assessment of Water Quality. Am. Soc. Test. Mater., Spec. Tech. Publ. 528: 117-126.

Craddock, D. R. 1977. Acute toxic effects of petroleum on arctic and subarctic marine organisms. In. Effects of Petroleum on Arctic and Subarctic Marine Environments and Organisms Ed. D.C. Malens Pub. Academic Press 2: 1-94.

Dicks, B. 1976. The importance of behavioral patterns in toxicity testing and ecological prediction. In Marine Ecology and Oil Pollution. Ed. J. Baker. Pub. J. Wiley & Sons. : 303-320

Hadley, D. 1977. Inter-and-Interspecific variability in tolerance of Southern California Littorina planaxis and Littorina scutulata to Petroleum. Environmental Research No. 3 pp. 186-208.

Kanter, R. 1974. Susceptibility to crude oil with respect to size, season and geographic location in Mytilus californianus (Bivalvia). Pub. University of Southern California, Sea Grant Program, Los Angeles, Ca. 90007. USC-SG-4-74, pp. 42.

Stein, R. J., Gundlach, E. R., Hayes, M. O. 1978. The Urquiola oil spill (5/12/76): Observations of biological damage along the Spanish coast. Conference on Assessment of Ecological Impact of Oil Spills, Keystone Colorado, June 1978.

Straughan, D. 1972. Factors causing environmental changes after an oil spill. J. Pet. Tech. Offshore Issue : 250-254.

Straughan, D. 1976. Sublethal effects of natural chronic exposure to petroleum in the marine environment. American Petroleum Institute Publication No. 4280. pp. 123.

Straughan, D. 1977. Biological survey of intertidal areas in the Straits of Magellan in January, 1975, five months after the Metula oil spill. Proc. Sym. Fate and Effects of Petroleum Hydrocarbons in Marine Ecosystems and Organisms. Sponsored NOAA, EPA. Seattle, November 1976: 247-260.

Straughan, D. and Hadley, D. 1978. Experiments with Littorina species to determine the relevancy of oil spill data from southern California to the Gulf of Alaska. Marine Environmental Research, 2.

Thomas, M. L. H. 1977. Long term biological effects of Bunker C Oil in the intertidal zone. Proc. Sym. Fate and Effects of Petroleum Hydrocarbons in Marine Ecosystems and Organisms. Sponsored NOAA EPA, Seattle, November 1976: 238-246.

10

THE USE OF GC/MS IN THE IDENTIFICATION AND ANALYSIS OF ORGANIC POLLUTANTS IN WATER

by

R.E. Finnigan
Finnigan Corporation

The growing concern over the many potentially hazardous chemicals being released in our environment has spurred the development of specific and sensitive methods for their analysis. Until recently, gas chromatography, with both specific and univeral detectors, has been the method of choice for the analysis of organic pollutants. However, with the myriad of structurally similar compounds recently discovered in water, air, and other environmental samples, positive identification has become difficult. In addition, the high number of contaminants in samples has created problems in the detection of ultra trace concentrations of pollutants.

The application of GC/MS to environmental pollution has solved both the sensitivity and identification problems. When operated in the normal scanning mode, a gas chromatograph/mass spectrometer provides valuable structural information and sensitivity equivalent to a standard flame izonization detector. When operated in the selected ion monitoring mode (also known popularly as the mass fragmentography mode), a GC/MS provides very specific information at a sensitivity level that often surpasses that of the electron capture detector. In the selective ion monitoring mode of operation, specificity and quantitative analysis are obtained by monitoring two or more ion fragments in combination with the gas chromatographic retention time.

GC/MS Technique for Determination of Interferences in Pesticide Analysis

The electron capture detector has long been employed in the analysis of chlorinated hydrocarbon pesticides. Its high sensitivity and excellent selectivity have made it the detector of choice in pesticide analysis. Some of the recent chemicals found in the environment are the polychlorinated bipheyls (PCB's). These industrial compounds are structurally similar to the chlorinated hydrocarbon pesticides and, thus, may interfere with qualitative and quantitative electron capture analysis of the pesticides.

PCB's and chlorinated naphthalenes have been separated by silicic acid column chromatography prior to analysis by gas chromatography (1,2). However, the use of gas chromatograph/mass spectrometer enables the qualitative and quantitative analysis of these compounds without prior column chromatography or complete separation on a gas chromatographic column. Goerlitz and Law (3) have recently analyzed chlorinated naphthalenes by GC/MS; however, they were not analyzed in combination with interfering substances.

Experimental

Apparatus. The instrument used for this analysis was a standard Finnigan Model 1015 GC/MS system which utilizes the quadrupole-type mass filter and glass jet separator interfacing the GC to the MS. A standard Finnigan data system was used to control the GC/MS and to provide data reduction. A Finnigan Programmable Multiple Ion Monitor (PROMIM) was used to perform the selected ion monitoring (SIM).

120

Material. Aroclor 1254 (polychlorinated biphenyls) is from Monsanto Chemical Company. The Halowax 1014 (chlorinated naphthalenes) is from Koppers, Industrial Chemicals Division, Verona, Pa. The pesticides were obtained from the Perrine Primate Research Branch, Environmental Protection Agency, Perrine, Fla.

GC Conditions. A 5-ft x 1/4-in. (2 mm i.d.) borosilicate glass column packed with 5% OV-17 on 6-/80 Gas Chrom Q was employed for the analysis of the chlorinated hydrocarbon pesticide mixture with the Halowax 1914 and Aroclor 1254. Temperature was programmed from 180-250° C and 8° C/min. The p,p'-DDE/p, p'-DDT/PCB mixture was chromatographed on a 5-ft x 1/4-in. (2 mm i.d.) borosilicate glass column packed with 3% OV-1 on 60/80 Gas Chrom Q at 200° C.

Results and Discussion

Qualitative Analysis. A standard chlorinated hydrocarbon pesticide mixture (Figure 1) was mixed 1:1 with Aroclor 1254 and Halowax 1014 to obtain Figures 2 and 3, respectively. Figures 1 to 3 are computer reconstructed chromatograms; the summation (amplitude) of the ion intensities of each spectrum is plotted against the spectrum number. Each pesticide is at a concentration of 100 ng/µl. The concentration of the Halowax 1014 and Aroclor 1254 is also 100 ng/µl. One-microliter samples were injected into the GC/MS.

In the Aroclor 1254 pesticide mix (Figure 2), there are several overlapping peaks in the region of spectrum numbers 84-94. Figures 4 (spectrum 85), 5 (spectrum 88), and 6 (spectrum 92) are computer plots of mass spectra stored on a magnetic disc. They were identified as hexachlorobiphenyl (m/e 358), p, p'-DDD (m/e 318), heptachlorobiphenyl (m/e 392), respectively. Note that there are very few interfering spectra in the three mass spectra of interest.

In the pesticide-Halowax 1014 mixture (Figure 3), the area chosen to study was between spectrum numbers 75 and 85. Figure 7 (spectrum number 78) was identified as dieldrin (m/e 378) while Figure 8 (spectrum number 82) was identified as an isomer of pentachloronaphthalene (m/e 298).

Quantitative Analysis. Using either the Finnigan PROMIN (Programmable Multiple Ion Monitor) or the data system, we can selectively monitor only those peaks of interest rather than the entire mass spectrum as in the previous analyses. The technique, known as a selective ion monitoring (SIM) provides several advantages over conventional full spectrum scanning: (1) Sensitivity can be increased by 100

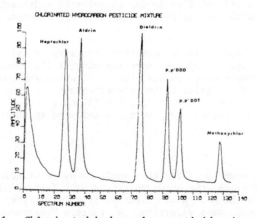

Figure 1.   Chlorinated hydrocarbon pesticide mixture

121

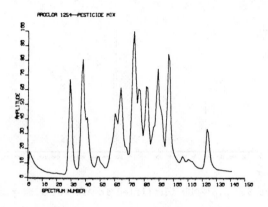

Figure 2. Aroclor 1254-pesticide mix

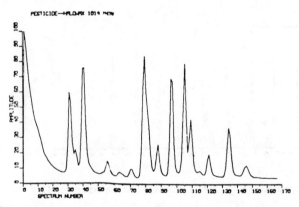

Fig. 3. Pesticide - Halowax 1014 mix

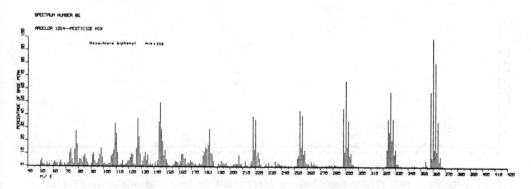

Figure 4. Hexachloro biphenyl identified in Aroclor 1254-pesticide mix

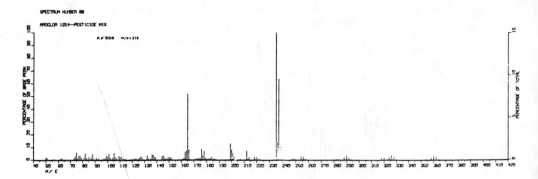

Figure 5. p,p'-DDD identified in Aroclor 1254-pesticide mix

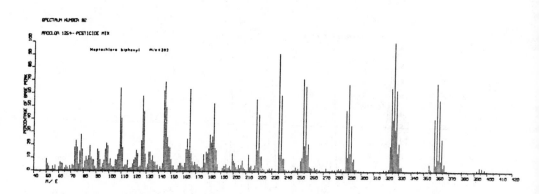

Figure 6. Heptachloro biphenyl identified in Aroclor 1254 pesticide mix

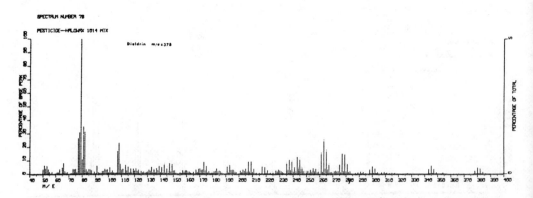

Figure 7. Pentachloro naphthalene identified in Pesticide-Halowax 1014 mix

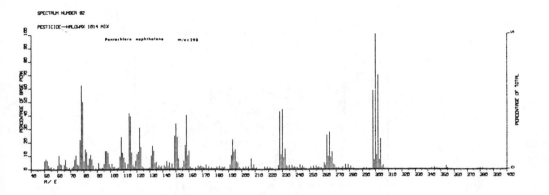

Figure 8. Pentachloro naphthalene identified in Pesticide-Halowax 1014 mix

to 500 times or more, (2) Interfering compounds or contaminants can be easily dis-
criminated against provided that they do not produce peaks at the same assigned
masses as the compound we wish to detect, (3) Accurate quantitative analyses can
be performed using appropriate standards.

Figure 9 illustrates the SIM technique in its simplest form, comparing the total ion
scanning mode in the analysis of the pesticide parathion with the single ion scanning
mode, for the same compound. In the latter case, we have chosen to monitor only the
base peak (m/e 109) of parathion. The Total Ion Monitor signal is shown as the top
recording in Figure 9, whereas the Single Ion Monitor output is shown on the lower
portion of Figure 9. The latter signal is many times greater than the corresponding
TIM because the quadrupole filter is controlled to transmit ions of only m/e 109
throughout the period that parathion elutes through the GC and into the MS. In the
full scanning mode, by contrast the mass filter transmits ions of m/e 109 only .01
seconds during each 3 second scan.

Figure 10 shows the selective ion monitoring of the base peak and molecular ion in
p, p'-DDE and p,p'-DDT, respectively, which were added to a sample containing PCB's.
The PCB's did not interfere with the analysis. Identification of a compound is
achieved by comparing its retention time to the presence of up to eight of its
characteristic mass peaks at the retention time. Quantitation is achieved by comparing the area under one or more of the eight peaks to the area of a known amount
of a standard. By SIM, sensitivities in the low nanogram and picogram range can
be obtained. Figure 11 is a typical linearity curve obtained by monitoring the
m/e 246 peak of p, p'-DDE; the sample size is that amount injected on column and
shows the sensitivity obtainable with this technique.

Conclusions
Components of overlapping gas chromatographic peaks in the 10 ng or less sample size
can be identified by the mass spectrum obtained. More recent GC/MS systems permit
identification with sample sizes of only 100 picograms. It is desirable to employ
a computer which has the capacity to store the entire analysis, process the date
(such as subtract background), and present it in a form which is easy to interpret.
Components of overlapping as chromatographic peaks present in amounts less than 10
ng can be identified and quantitated by the technique of selective ion monitoring.

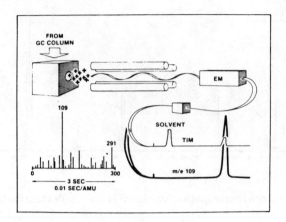

Fig. 9.   Single and total ion scanning.

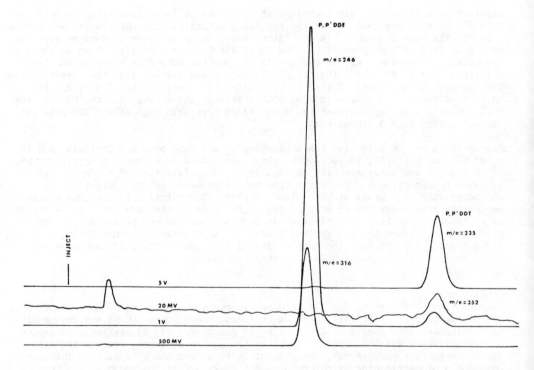

Fig. 10.   Peak monitoring for p, p'-DDE and p, p'-DDT in a PCB mixture.

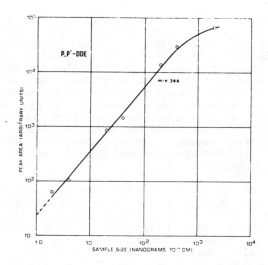

Fig. 11. Linearity curve monitoring the m/e 246 peak of p, p'-DDE

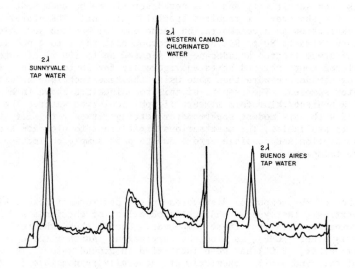

Fig. 12. Chloroform analysis in various water samples

126

Analysis of Hazardous Environmental Chemicals by Selective Ion Monitoring

Organics in Water

In recent years, there has been a concern over chloroform and other galogenated hydrocarbons detected in municipal water supplies. Although the highest concentrations of these compounds found were in the 37-150 g/1 (ppb) concentration range, it has been determined that these do not represent an acute hazard to man at this level. The compounds form as a result of chlorination processes during water treatment. The addition of chlorine at various stages of treatment is the determining factor in the final amount of halogenated compounds found in drinking water. Considering the widespread use of chlorine in water, sewage treatment processes, household and commercial laundering, paper pulp bleaching, and related processes, it is possible to postulate the widespread occurrence of these compounds. Some of the organochlorine compounds found are methylene chloride, chloroform, 1, 1,1-trichloroethane, 1,1,2-trichloroethylene, 1,1,2,2-tetrachloroethylene, dichlorobenzenes, and trichlorobenzenes.

The technique of direct aqueous injection described by Harris, Budde, and Eichelberger of the EPA (4) was employed in our work. Stock solutions could be kept 3 to 4 days when tightly sealed and refrigerated; however, dilutions were made every day. The standard samples were analyzed along with water samples from California, New Mexico, Canada, and Argentina (Figure 12). A graph (Figure 13) was prepared from these results, and was used to calculate the chloroform concentrations (Buenos Aires river, 3.5 ppb; Buenos Aires tap water, 20.5 ppb; Carlsbad tap water, 20 ppb; Sunnyvale tap water, 35 ppb; Western Canada Chlorinated, 50 ppb; chemical spill, 90 ppb). Table 1 compares the calculated ration of m/e 83 to m/e 85 in chloroform and m/e 117 to m/e 119 in carbon tetrachloride to the observed ratios at various concentration levels injected into the GC/MS. The observed ratios give the permissible limits in which to analyze unknowns, and conclude that they are free of coeluting impurities.

Hexadecane Extraction

The Bellar Lichtenberg sparge and trap method (5) for analysis of volatile organics in water has good sensitivity and does not discriminate between most compounds of varying polarity; however, it requires special equipment. Therefore, we have used hexadecane extraction as a concentration technique for benzene and chloroform. This procedure extracts 80 to 95% of this compound at the 1 to 4 ppb level. One milliliter of steam-treated hexadecane was floated on to 100 ml of water sample in a tightly closed container and agitated vigorously for 15 to 20 min. Two microliter injections of hexadecane were then made using the same technique as with the direct injected water samples. The results of this technique are shown in Figure 14 of the analysis of a benzene/chloroform mixture (1 ppb each) from water. The SIM's shown were obtained with very modest spectrometer settings; twenty to fifty times more sensitivity is available. The most serious limitations to ultimate sensitivity are GC column absorption and possible errors due to poor sample collection, storage, and shipping technique.

Carcinogens

A large number of the compounds discovered in the environment have exhibited carcinogenic properties. Last year, OSHA published a list of chemicals and has established rules to protect workers from these compounds. One of these chemicals, bis (chloromethyl) ether (BCME), is a potent carcinogen. Another toxic compound, bis (chloroethyl) ether, (BCCE) has been identified in ground water. Industrial plant effluents of this and similar chemicals are generally responsible for their identification in ground water and river water. Figure 15 is a mass spectrum of BCEE. The ions monitored were m/e 93 and 95, and are shown in Figure 16, along with the

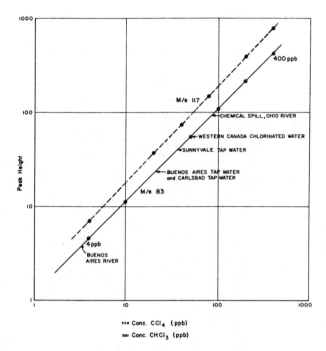

Fig. 13. Standard curve for CHCl$_3$ and CCl$_4$ in water.

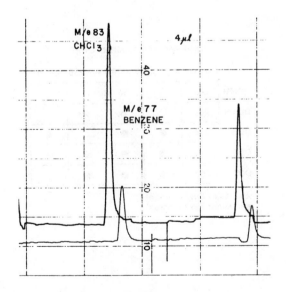

Fig. 14. Two channel programmable multiple ion monitor chromatogram of principal ion fragments of chloroform and benzene.

TABLE 1

ION FRAGMENT RATIOS

ppb	$CHCl_3$ M/e $83/85$ = 0.654	CCl_4 M/e $117/119$ = 0.981
800	—	.98 ± .003
400	.65 ± .005	.98 ± .003
200	.66 ± .005	.98 ± .003
100	.65 ± .005	—
80	—	.96 ± .01
50	.66 ± .02	—
40	—	.99 ± .01
20	—	.95 ± .03
10	.63 ± .02	—
4	.67 ± .03	.94 ± .03

All RATIOS ARE AN AVERAGE OF 3 TO 5 DATA POINTS

TABLE 2

ION FRAGMENTS MONITORED

	Mol. Wt.	M/e 282	M/e 284	M/e 235	M/e 318
DDMU	282	★	★		
DDMS	284		●	★	
DDE	316	●	●		★
DDD	318	●	●	★	●
DDT	352	●	●	★	

★ = Major Ion Fragments
● = Minor Ion Fragments

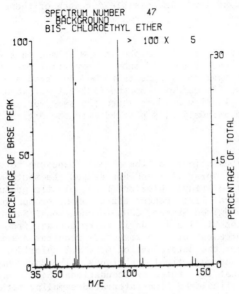

Fig. 15. Mass spectrum of bis (chloroethyl) ether.

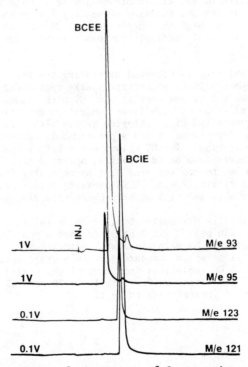

Fig. 16. Four-channel mass fragmentogram of fragment ions in BCEE and BCIE.

major ion fragments of a related compound, bis(chloroisopropyl) ether (BCIE).
Quantitation at the 10-pg level is possible for each of these compounds.

Whole Fish Extract

A search for DDT and its principle metabolites was made in a whole fish extract provided by a federal agency. Figure 17 shows the partial mass spectra of DDT and its metabolites. The principal ion fragments to be monitored are summarized in Table 2. Figure 18 shows the SIM output from the fish extract when using the PROMIN and strip chart recorder. In Figure 19 is shown the same data when the SIM is generated by the data system and automatically plotted with the digital data plotter.

Vinyl Chloride

Recently, the deleterious effects of the chronic exposure to vinyl chloride has led to the recall of aerosol spray cans and has caused the shutdown of a plant in Europe. As a starting material polyvinyl chloride (PVC), plastic production vinyl chloride has been linked to several liver cancer deaths among production workers. Beginning January 1, 1975 in the United States, OSHA (Occupational Safety and Health Administration) has reduced the vinyl chloride in workplace air from 50 to 1 ppm over an 8-hr period. The enforcement of such standards requires a sensitive and specific analytical technique for the determination of vinyl chloride. Such a method must include both sampling and analytical techniques. The Environmental Protection Agency (EPA) and National Institutes of Occupational Safety and Health (NIOSH) have developed similar analytical and time-integrated sampling methods. The vinyl chloride sampling procedure used in this work involved drawing a known volume of air (10 μl) through two-section charcoal trap by means of a small calibrated air pump. The front section of the trap contains 100 mg and the rear section 50 mg of charcoal. The two portions are analyzed separately. When the vinyl chloride content of the sampled air is high and when more than 20% of the total vinyl chloride is found in the rear charcoal section, the possibility of trap overload and sample loss exists.

Desorption of vinyl chloride is effected by placing the head and back of the charcoal adsorbent in separate 50-ml volumetric flasks containing 30 ml of cold (-20° C) carbon disulfide. The flasks are kept at -20° C, with frequent swirling, for 1 hr. NIOSH and EPA have relied on GC with flame ionization detection with confirmation by GC/MS. We have quantified vinyl chloride by SIM with a Finnigan Model 3200F gas chromatograph/mass spectrometer and a programmable multiple ion monitor (PROMIN) for selective ion monitoring. The GC column was a 5-ft x 2-mm-i.d. glass column, packed with 0.4% Carbowax 1500 on Carbopack A, operated at 35° C. Two of the PROMIN channels were set to monitor m/e/ 62 (the molecular ion of vinyl chloride) at 10X different sensitivity levels. This provided a 10 times wider dynamic range than is usually obtained with a strip-chart recorder.

Peak height reproducibility was quite good, as shown in Figure 20. Six consecutive injections of 100 μl (250 pg) of vinyl chloride gave a peak height of 122 mm + 2 mm (1.6%). Such precision can be maintained with solution standards. Quadruplicate analysis of 0.06 ng/ μl solution standard gave an average deflection of 58.2 mm + 0.2 mm/3 μl injection. CS$_2$ solutions from control charcoal filters had no detectable vinyl chloride. The detection limit is 10 to 20 pg/2 μl injection with a signal-to-noise ratio of greater than 10 to 1.

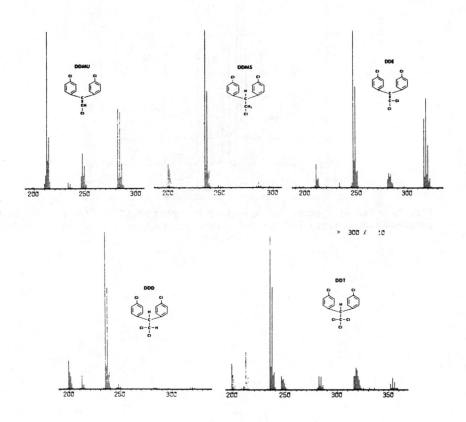

Fig. 17. Partial mass spectra of DDT and its metabolites.

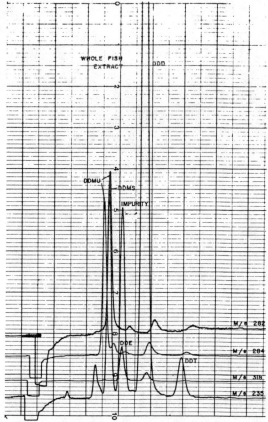

Fig. 18. Manually generated mass fragmentogram of whole fish extract using programmable multiple ion monitor. Pens are mechancially offset to reduce peak overlap.

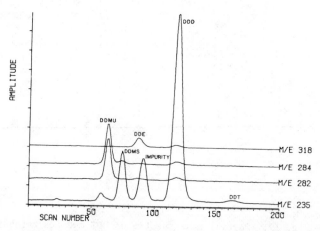

Fig. 19. Computer generated mass fragmentogram of whole fish extract.

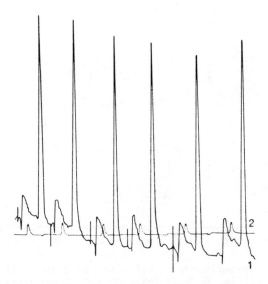

Fig. 20. Repetitive injections of 100ml of 1-ppm vinyl chloride in air.
Channel 1 is about 10 times more sensitive than channel 2. Injection point
is marked by a vertical spike.

High Resolution Capillary Column GC/MS

This section presents data which show the value of high resolution glass capillary
column GC used in combination with both electron impact (EI) and chemical ionization (CI) quadrupole mass spectrometry for analysis of organic compounds which are
of interest to environmental chemists. Often, these techniques provide detection
and identification at very low levels of compounds which do not lend themselves to
analysis by conventional packed column GC/MS using electron impact ionization alone.

Polycyclic Aromatic Hydrocarbons in Crude Oil

As an example of the power of high resolution capillary columns in GC/MS, we show
the analysis of an Alaskan crude oil sample. One of the major objectives was to
determine if polycyclic aromatic hydrocarbons (PAH) were present in the crude oil
samples, and their level. It is suspected that the PAH compounds are concentrated
in the refining process and could possibly reach dangerous levels.

Because of the rapid elution of many of the hydrocarbon peaks from GC, it is essential that the mass spectrometer be scanned very rapidly (i.e., 1 sec. or less)
when analyzing samples of this type. This is necessary to retain the resolution
achieved by the GC column.

Figure 21 shows the difference in the Total Ion Chromatogram obtained when using 1
sec. and 3 sec. scans in the analysis of an oil sample by capillary column. As
can be seen, the GC resolution is seriously degraded when scanning at the 3 sec./
scan rate. The ability of the quadrupole mass spectrometer to scan at high speed,
repeatedly, is a necessary feature for capillary GC/MS analysis.

134

OIL SAMPLE BY GLASS CAPILLARY EI

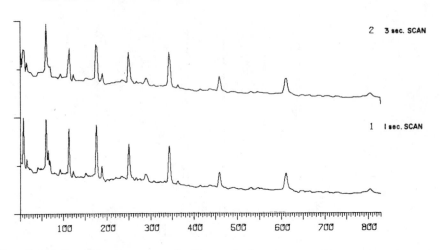

Fig. 21.   Total Ion Chromatograms of oil sample by glass capillary GC and El-MS,
when scanning at 1 sec./mass scan and 3 sec./mass scan.

We show in Figure 22 the conditions used for the glass capillary column EI-MS analysis of the Alaskan oil sample. In this example, EI-MS is used rather than CI-MS,
since the PAH compounds lend themselves to EI interpretation.

The results of the analysis are shown in the following figures. Figure 23 shows
the Total Ion Chromatogram, indicating the location of components of interest. The
m/e values marked in the chromatogram correspond to spectra to where specific ion
searches indicated these masses present. The m/e values searched are specific for
PAH compounds. A total of 1825 scans were taken over a 45-minute period.

Note also that compounds with MW at m/e = 252, which very likely indicate where the
most carcinogenic PAH might occur, are present at very low levels as compared to
other components. Figure 24 shows a table of polynuclear aromatic hydrocarbons
arranged by molecular weight. The ones labeled (*) are known carcinogens (reference
Merck index).

Figure 25 shows a specific ion search for m/e 178 (possibly anthracene and phenanthrene) overlapped with the Total Ion Chromatogram. Figure 26 shows the mass
spectrum for either anthracene or phenanthrene (MW - 178); both give essentially
identical mass spectra. Either standards or exact relative retention time (RRT)
are needed to determine which compound is being identified. Figure 27 shows the
mass spectrum #999, a PAH with MW = 228. Again, either a standard or accurate RRT
is needed to determine exact identity. Similarly, Figures 28 and 29 show mass
spectra of scan numbers 1163 and 1233 respectively. Both show PAH compounds with
MW = 252, those PAH most often associated with carcinogenic activity. Note the
presence of strong doubly charged ion in the region of m/e 126.

When analyzing and identifying normal paraffins in an oil sample such as this,
methane CI provides significant molecular weight information. The EI spectrum of
a normal paraffin is shown in Figure 30 whereas the CI spectrum of another

CONDITIONS USED FOR ANALYSIS OF

ALASKA OIL SAMPLE BY GLASS CAPILLARY

GC - EI - MS

Column:	20 meter x 0.02 inch ID
	SE-30 glass capillary
Column Temperature:	Ambient to 250°C at 4°C/min.
Carrier Gas:	Helium at 2cc/min flow
Injector Temperature:	250°C
Sample Size:	0.2 microliter (Splitless injection)
Mass Range Scanned:	40 - 350 amu
Scan Rate:	1.227 seconds each

Fig. 22.   Conditions employed from analysis of Alaskan oil sample.

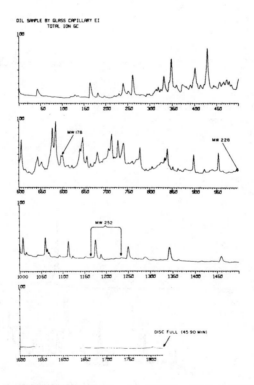

Fig. 23.   Total Ion Chromatogram of Alaskan oil sample.   The m/e values
marked on the chromatograms show where specific PAH compounds were
found by limited mass search.

POLYAROMATIC HYDROCARBONS THAT PRODUCE VERY SIMILAR EI
MASS SPECTRA - BUT ARE IDENTIFIABLE BY CAPILLARY GC-MS

$C_{20}H_{12}$ MW 252

1. 3,4-benzfluoranthene

2. 3,4-benzpyrene   *

3. 1,2-benzpyrene   *

4. 10,11-benzfluoranthene

5. 11,12-benzfluoranthene

6. Perylene   *

_____

*
     Known carcinogens

$C_{14}H_{10}$ MW 178

1. Phenanthrene   *

2. Anthracene

$C_{18}H_{12}$ MW 228

1. Triphenylene

2. 3,4-benzophenanthrene

3. 1,2-benzanthracene

4. Chrysene

5. Napthacene

Fig. 24. Table of polynuclear aromatic hydrocarbons.

OIL SAMPLE BY GLASS CAPILLARY EI
1 IS TIC.2 IS M/E 178

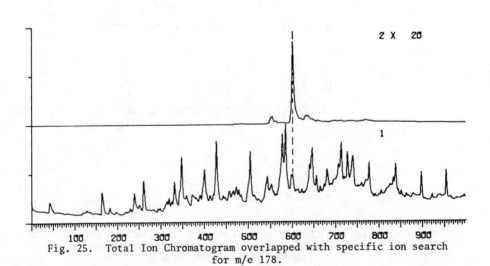

Fig. 25. Total Ion Chromatogram overlapped with specific ion search
for m/e 178.

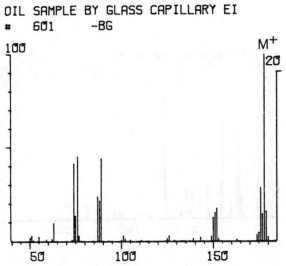

Fig. 26. Mass spectrum for either anthracene or phenanthrene.

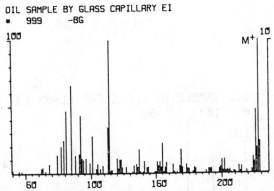

Fig. 27. Mass spectrum of scan #999 which represents PAH compound with MW≑228.

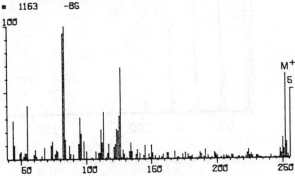

Fig. 28. Mass spectrum of scan #1163 which represents PAH compound with MW=252.

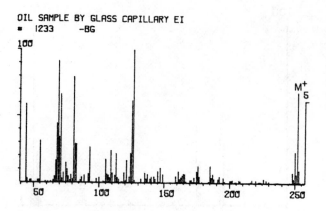

Fig. 29. Mass spectrum of scan #1233 which represents PAH compound
with MW = 252.

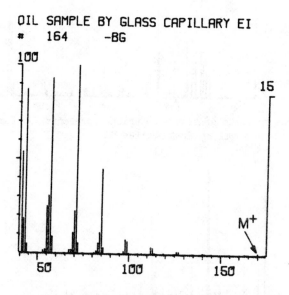

Fig. 30. EI spectrum of normal paraffin with molecular ion at m/e 170.

homologous compound is shown in Figure 31. The M-1 ion is very strong in the methane CI spectrum as compared to the very small molecular ion at m/e 170 in the EI spectrum.

We next show the results of an analysis performed on a sediment extract taken from Puget Sound near Seattle, Washington, heretofore considered a pristine area of the United States. The extract was taken from a 100 gram sediment sample in an area which had suspected exposure to No. 2 fuel oil.

A 1- μl injection onto the Finnigan Model 3200 GC/MS was made from a 0.5 ml final extract volume. A 20 meter SE-30 glass capillary column directly interfaced to the MS ion source with an all glass capillary section was used for this analysis. It was programmed from 60° C to 250° C at 6° C/minute after 3 minutes inital hold. The scan rate was 1.3 seconds for the mass range of interest (80 to 280 amu). An internal standard was used for quantitative purposes. This was hexamethyl benzene at a concentration of 8 ng/ μl.

Figure 32 shows the Total Ion Chromatogram which is alternately referred to as the Reconstructed Gas Chromatogram (RGC). In addition, we have shown limited mass chromatograms for m/e 57 and m/e 71. Figure 33 shows the RGC for a PAH standard run on the same column under the same conditions as above. Figure 34 shows limited mass searches for a series of ions as well as the RGC, which serve to identify the compounds of interest. Table 3 tabulates the identities of each of the components numbered on Figure 34.

We are indebted to Mr. Randy Jenkins of National Oceanographic and Atmospheric Administration, who performed the above analysis and generously supplied all of the data.

Sterols by Capillary Column GC/EI/MS

The sterols considered here are often encountered when natural products are being investigated. The methyl ether derivatives make these compounds more suitable for GC analysis than the corresponding free alcohols. Often these components are present at very low levels in highly complex mixtures.

Figure 35 shows the structure of three of the sterols considered here: chloresterol methyl ether (MW = 400); 24-methyl-cholesterol methyl ether (MW = 414) and 24-methylene cholesterol methyl ether (MW = 398). Figure 36 shows the strucuture of stigmasterol methyl ether (MW = 426) and betasitosterol methyl ether (MW = 428).

We analyzed a water sample for sterols sent to us by the Canada Centre for Inland Waters using packed column GC/MS. It is interesting to compare the packed column results with those obtained by capillary column GC/MS. Figure 37 shows the Total Ion Chromatogram obtained by GC/MS of an extract taken from the Canadian waters where packed column GC techniques were used to attempt to identify sterols in water. It is believed that the sterol analysis was to determine if raw sewage was being dumped into the water since these are known fecal sterols. There is no baseline separation between stigmasterol and beta-sitosterol (peaks 6 and 8).

Next we analyzed a known synthetic mixture of the sterol methyl ethers shown in Figure 35 and 36 by high resolution capillary column GC together with high sensitivity, fast scanning MS (Finnigan Model 3200). The sample was provided through the generosity of Stanford University Chemistry Department. A coaxial interface was used between the GC and the MS, as shown in Figure 38. This provides EI operation when the reactant gas is shut off and CI operation when the reactant gas is added at point 2. Figure 39 shows the conditions used to analyze the sterol mixture. Note that the scan time used was 2 seconds for the mass range of interest.

140

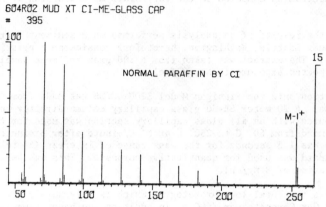

604R02 MUD XT CI-ME-GLASS CAP
395

100

NORMAL PARAFFIN BY CI

15

M-1⁺

50   100   150   200   250

Fig. 31.   Methane CI spectrum of normal paraffin.

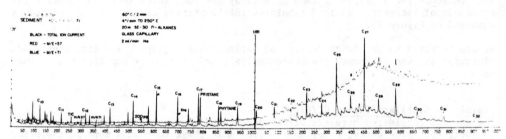

Fig. 32.   Total Ion Chromatogram and Limited Mass Chromatograms (m/e 57,71)
obtained from Puget Sound sediment extract.

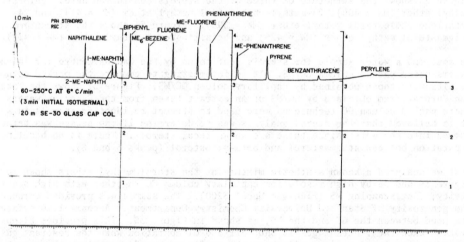

Fig. 33.   Total Ion Chromatogram for polycyclic aromatic hydrocarbon (PAH) standard.

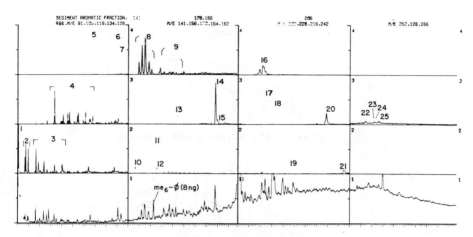

Fig. 34. Total Ion Chromatogram and Limited Mass Chromatograms obtained from Puget Sound sediment extract.

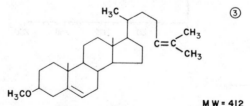

CHOLESTEROL METHYL ETHER   MW = 400

24-METHYLCHOLESTEROL METHYL ETHER   MW = 414

24-METHYLENECHOLESTEROL METHYL ETHER   MW = 412

Fig. 35. Molecular structure of three sterol methyl ethers analyzed by capillary column GC/MS.

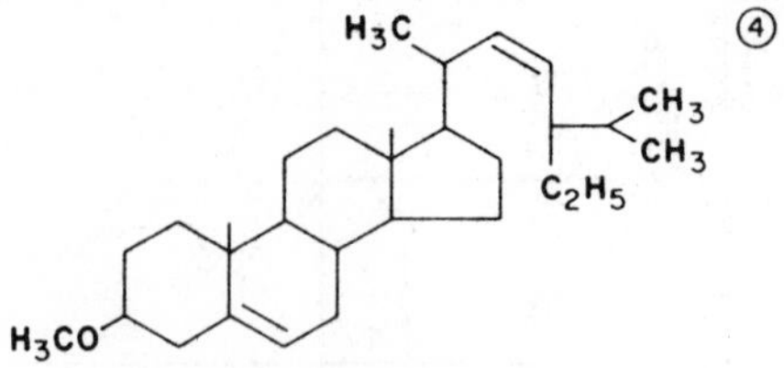

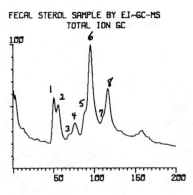

④

STIGMASTEROL METHYL ETHER M W = 426

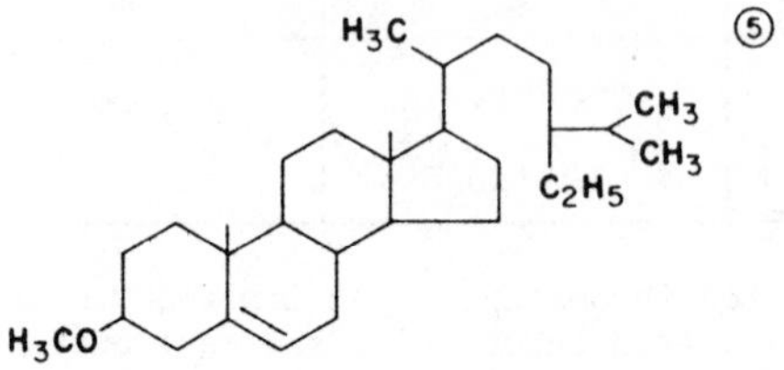

⑤

β - SITOSTEROL METHYL ETHER M W = 428

Fig. 36. Molecular structure of stigmasterol methyl ether
and beta-sitosterol methyl ether.

FECAL STEROL SAMPLE BY EI-GC-MS
TOTAL ION GC

Fig. 37. Total Ion Chromatogram of fecal sterol sample taken from
Canadian waters. Analysis was performed using packed column GC/MS.

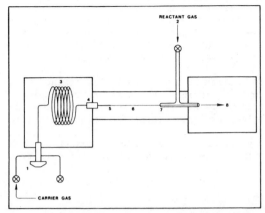

Fig. 38. Coaxial dual-gas interface used to interface glass capillary
column to MS for EI and CI-MS. (1) Gas Chromatograph with Grob-type
injector. (2) Reactant gas line (stainless steel, OD 1.6 mm, ID 0.5mm).
(3) Glass capillary column (H. and G. Jaeggi, Trogen, Switzerland).
(4) Polyimid swagelok (Vespel SP-1, Dupont) on Teflon shrink-tube.
(5) Glass/platinum seal. (6) Interface capillary (platinum, OD 0.3mm,
ID 0.1 mm). Coaxial Dual-gas interface. (8) Finnigan Model 3200 quadrupole mass spectrometer.

STEROL GC/MS ANALYSIS CONDITIONS

Column:	10 meter x 0.02 inch ID
	SE-30 glass capillary
Column Temperature:	$265^{o}C$ isothermal
Carrier Gas:	Helium at 2 cc/min flow
Injector Temperature:	$265^{o}C$
Split Ratio (Injector):	10 to 1
Sample Size:	200 nanograms
Ionization:	Electron impact
Mass Range:	50-550
Scan Time:	2.0 seconds

Fig. 39. Conditions used to analyze sterol mixture by capillary column GC/MS.

144

Figure 40 shows the resulting capillary Total Ion Chromatogram. The analysis took less than 20 minutes (as compared to approximately 6 minutes in the prev`ous example using a packed column). Component 4 represents 20 ng of stigmasterol methyl ether. There is a significant improvement in resolution between stigmasterol and beta-sitosterol (peak 4 and 5). The background subtracted EI mass spectrum of stigmasterol (MW =426) is shown in Figure 41.

Figure 42 shows the EI spectrum of beta-sitosterol obtained from the Canadian waters fecal sterol sample. Figure 43 shows the corresponding CI-methane spectrum of beta-sitosterol for comparison. The CI spectrum adds considerable information in the region of the molecular ion, helpful for unique identification.

It is concluded that high resolution capillary column GC in combination with high sensitivity, fast scanning MS is ideal for analyses of this type.

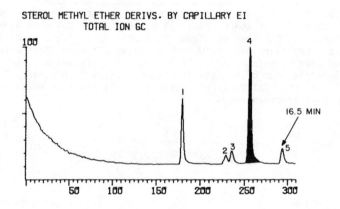

Fig. 40. Total Ion Chromatogram of sterol methyl ether mixture analyzed by capillary column EI-MS.

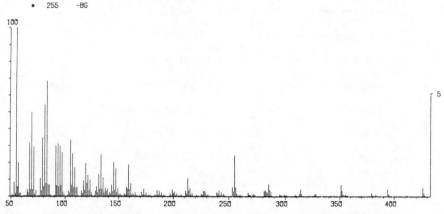

Fig. 41. Mass spectrum of stigmasterol methyl ether.

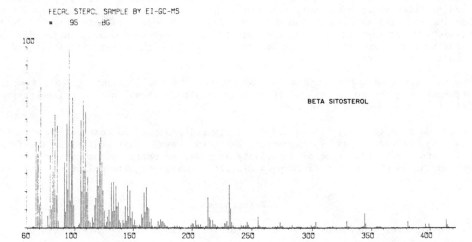

Fig. 42. EI mass spectrum of beta-sitosterol in fecal sterol sample taken from Canadian waters.

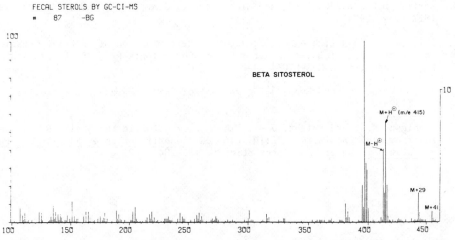

Fig. 43. CI mass spectrum of beta-sitosterol in fecal sterol sample taken from Canadian waters.

Conclusion

It has been shown that GC/MS provides the most powerful and advanced technique available today for the solution of environmental problems. When used in the full scanning mode, it provides sensitivity at least equivalent to that of the gas chromatograph equipped with FID, but provides as well valuable structural information which permits unambiguous identification of all compounds detected. In most instances, the GC/MS data system can assist in this identification using vast libraries which have now been compiled.

146

When the GC/MS system is used in the selected Ion Monitoring mode, it provides sensitivities which often surpass those of a GC equipped with electron capture, yet has the specificity to unambiguously identify the compound of interest while discriminating against interfering contaminants and other compounds. In the SIM mode, quantitative measurements can be performed at ultra-high sensitivities with accuracies of $\pm 3\%$, or better.

The most advanced GC/MS systems incorporate both EI and CI modes of ionization. The information gathered in each of these modes has been shown to be complementary when detecting and identifying unknowns.

The value of high resolution glass capillary columns used in combination with fast scanning quadrupole mass spectrometers has been shown. Most systems of the future will incorporate this type of capability in order to optimize the task of separation and identification of complex mixtures such as organic pollutants in water.

References

1. Amour, J.A. and Burke, J.A., J.Ass. Offic. Anal. Chem. 53 761-768 (1970).

2. Ibid., 54, 175-177 (1971).

3. Goerlitz, D. and Law, L., Bull. Environ. Contam. Toxicol. (in press).

4. Harris, L.E., Budde, W.L., and Eichelberger, J.W., Anal. Chem., 46(13), 1912-1917 (Nov. 1974).

5. Bellar, T.A. and Lichtenberg, J.J., "The Determination of Volatile Organic Compounds at the µg/l Level in Water by Gas Chromatography," EPA Environmental Monitoring Series, EPA-670/4-74-009.

6. Vander Velde, G. and Ryan, J.F., Journal of Chromatographic Science, 13, 322-329 (1975).

7. Finnigan, R.E. and Knight, J.B., "The Use of GC/MS in the Analysis of Unusual Environmental Chemicals," KEITH, L.A. ed. Identification & Analysis of Organic Pollutants in Water, Ann Arbor Science, 185-204 (1976).

11

DEVELOPMENT OF A NOVEL HYDROCARBON-IN-WATER MONITOR
I. Concept Feasibility Demonstration

by

H.S. Silvus, Jr.[1], F.M. Newman[1], G.E. Fodor[1], and F.K. Kawahara[2]
Southwest Research Institute

INTRODUCTION

Oil, at present, is one of the main sources of energy necessary to power our industries, to provide lubricants, and to yield many petroleum products which have innumerable uses in our everyday living. All of these are essential for our national well-being and security. However, accidents, negligence, and mechanical failures result in wasteful and harmful discharges of oil to the water environment. The largest source of oil pollution is associated with spill incidents involving transportation by tankers, river barges, pipelines, trucks, and rail tank cars. Since the huge volumes of crude and refined oil products transported by sea far exceed the combined tonnage of all other commodities, a fractional-percent spill of the total volume can be sizeable in cost, energy loss, and damage to the environment (Ref. 1).

As the discharge of oils into surface waters remains one of the serious and continuing water problems, monitoring of discharged petroleum products is an important necessity for energy and resource conservation and for environmental improvement. To provide monitoring for the waste-water treatment processes involved in coal liquefaction or shale-oil recovery, a suitable, inexpensive analytical instrument for quantitation of oil-in water is needed. Radar and microwave surveillance systems are sophisticated and expensive when compared to those devices involving ultraviolet or infrared absorption spectrophotometry. Moreover, before measurements can be made by these latter methods, the aqueous oil sanple must be extracted with a measured quantity of a toxic organic solvent, separated from the water phase, and dried, and then the oil must be quantitated by ultraviolet or infrared spectrometry.

To circumvent the above difficulties, a new concept in monitor design was investigated, and its feasibility for adoption into a "prototype" instrumental monitor was established (Refs. 2and 3).

TECHNICAL BACKGROUND

A. Optical Fiber Terminology and Properties

Optical fibers are usually made of two materials arranged coaxially as in Figure 1. The inside part of the fiber is referred to as the "core", and it can be made of plastic, glass, fused silica, sapphire, or in some cases, a liquid. Normally, the fiber core is surrounded by a layer of material which has lower index of refraction than the core and is referred to as the "cladding". Claddings are typically made

1. Southwest Research Institute, San Antonio, Texas
2. Environmental Monitoring and Support Laboratory, U.S. Environmental Protection Agency, Cincinnati, Ohio

of low-index plastic, glass, or silicone. A protective jacket of plastic or other material may be applied over the cladding to increase mechanical strength of the fiber and to facilitate handling of the fiber.

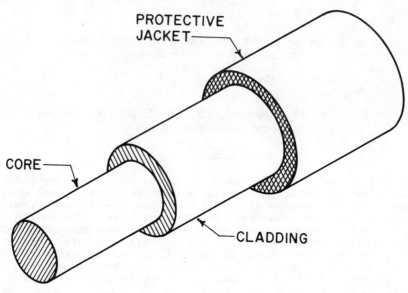

PROTECTIVE JACKET

CORE

CLADDING

Fig. 1. Construction of a Typical Clad Optical Fiber

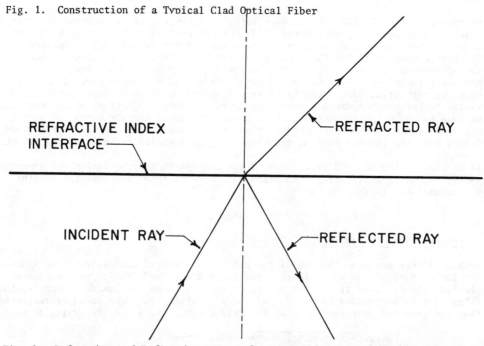

REFRACTIVE INDEX INTERFACE

REFRACTED RAY

INCIDENT RAY

REFLECTED RAY

Fig. 2. Refraction and Reflection at a Refractive Index Interface (Material above interface has lower index of refraction than material below interface.)

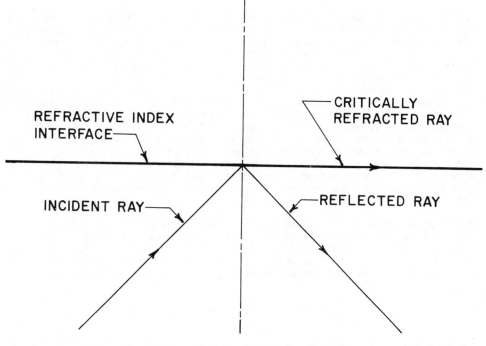

Fig. 3. Critical refraction at a Refractive Index Interface (material above interface has lower index of refraction than material below interface.)

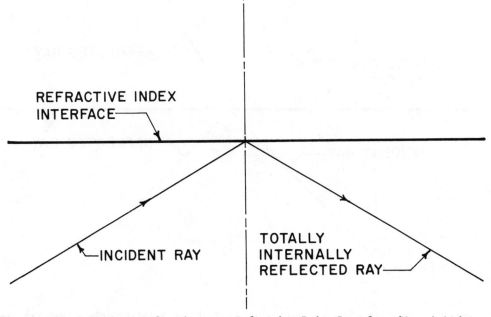

Fig. 4. Total Internal Reflection at a Refractive Index Interface (Material above
interface has greater index of refraction than material below interface.)

Refraction and reflection effects at a refractive-index interface are important to propagation of light through an optical fiber and to the system concept which is the subject of this paper; these effects will be reviewed briefly. Normally, when a light ray is incident on the boundary between two transparent materials of differingindex of refraction, a portion of the incident energy is reflected, and the remainder of the energy is refracted or bent through an angle as it enters the second material. In particular, when the refractive index of the material which the light ray is entering is less than that of the material which it is leaving, the refracted ray is bent toward the material boundary as shown in Figure 2. As the angle of incidence is increased, a critical angle is reached at which the refracted ray is transmitted along the material boundary as shown in Figure 3. As the angle of incidence is increased beyond the critical angle, the ray never enters the second material, but instead is deflected back into the material from which it came as shown in figure 4 by a phenomenon known as "total internal reflection". Light rays propagating through the core by total internal reflection at the core-cladding interface.

In the opposite case, i.e., when the refractive index of the material which the light ray is entering is greater than that of the material which it is leaving, the ray is bent away from the material boundary as in Figure 5. In this situation no critical angle exists, and an appreciable portion of the light energy incident upon the boundary is transmitted through the boundary as a refracted ray.

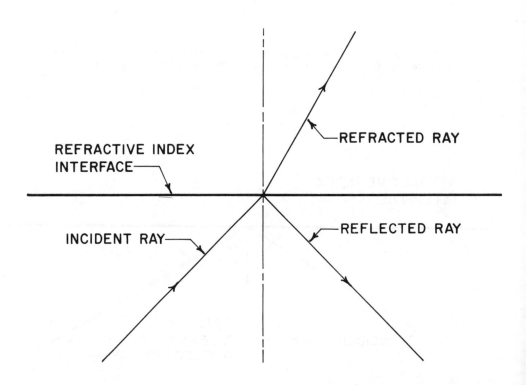

Fig. 5. Refraction and Reflection at a Refractive Index Interface (Material above interface has greater index of refraction than material below interface.)

B. System Concept

The reported hydrocarbon-in-water monitoring concept is based upon alteration of
the core-cladding interface of an optical fiber which in turn affects optical transmission through the fiber. A conventional clad optical fiber, such as that illustrated previously in Figure 1, is insensitive to the material in contact with the
outer surface of the cladding; hence, a clad optical fiber can be immersed in almost
any medium with essentially no effect on optical transmission properties of the
fiber. However, the optical transmission properties of an unclad fiber, that is,
one which has only a core and no cladding or protective jacket, are sensitive to
the medium in which the fiber is immersed since that medium effectively functions
as the fiber cladding. If the medium has lower index of refraction than that of
the fiber core material, then light traveling through the fiber is retained within
the fiber by total internal reflection and is transmitted with relatively low
attenuation to the opposite end of the fiber. On the other hand, if the surrounding medium has greater index of refraction than that of the core material, then
light traveling through the fiber is rapidly dissipated into the medium as illustrated in Figure 5 since total internal reflection does not occur.

In the reported hydrocarbon-in-water monitoring technique, an unclad optical fiber
is chemically treated to render its surface organophilic so that the surface of
the fiber will adsorb hydrocarbons. If an unclad optical fiber having greater
index of refraction than that of water is organophilically treated and immersed in
water, then water becomes the effective cladding. Since the index of refraction
of refraction of the water cladding is lower than that of the fiber core, total
internal reflection will occur, and light entering one end of the fiber will be
propagated to the opposite end of the fiber as shown in the left side of Figure 6.
However, if a small quantity of hydrocarbon material having index of refraction
greater than that of the optical fiber core material contaminates the fiber surface,
then total internal reflection will be degraded or destroyed in the area contacted
by the contaminant, and light propagating through the fiber will escape into the
medium as illustrated in the right side of Figure 6, thereby reducing the intensity
of light arriving at the output end of the fiber. The degree to which total internal reflection is degraded is related to the quantity of contaminant (i.e., thickness and extent of adsorbed layer) deposited on the fiber surface. Increasing

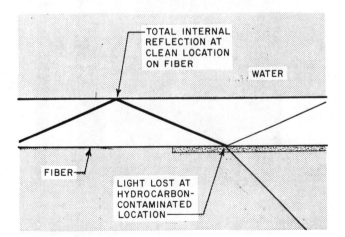

Fig. 6. Unclad Optical Fiber Immersed in Water Showing Light Reflection and Transmission Conditions at Clean and Hydrocarbon-Contaminated Locations

adsorbed layer thickness decreases the strength of total internal reflection and consequently increases the amount of light lost from the fiber core into the surrounding medium which, in turn, decreases the intensity of light arriving at the output end of the fiber.

Fig. 7. Optical fiber loop in air showing transmission of light over entire length.

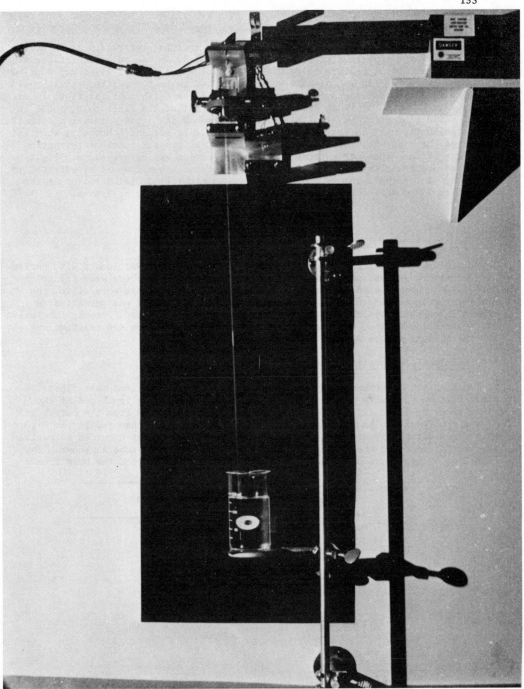

Fig. 8. Optical Fiber Loop Immersed in Liquid Showing Termination of Light Transmission by Loss of Total Internal Reflection

C. Illustration of System Concept

To illustrate the concept of destruction of total internal reflection in an unclad opticalfiber, the apparatus of Figure 7 was assembled. An unclad optical fiber was suspended from a test fixture to form a U-shaped loop, and laser light was coupled into one end of the fiber. A small weight was used to keep the fiber straight and to cause the bottom of the loop to sink in a fluid. With the optical fiber immersed in air, which had lower index of refraction than that of the fiber core, total internal reflection occurred, and the fiber appeared as in Figure 7 in which optical transmission through the entire length of the fiber is evident. However, when the bottom of the loop was immersed in a fluid which had greater index of refraction than that of the fiber core, total internal reflection was destroyed, and light ceased to propagate through the right-hand leg of the fiber loop as illustrated in Figure 8. This experiment is a dramatic, though simplified, illustration of the concept employed in the reported hydrocarbon-in-water monitoring system.

EXPERIMENTAL PROGRAM

A. Chemical Treatment Procedures

To accomplish chemical treatment, a special holding fixture was loaded with unclad optical fibers and was placed into a reaction vessel. The fibers were cleaned by successive washings with a solvent and water. After the fibers were clean, the treatment reagent was introduced into the reaction vessel and was permitted to remain in contact with the fibers for a period of several hours. At the conclusion of the reaction period, the treatment reagent was drained from the reaction vessel, and the optical fibers were washed and dried.

B. Experimental Apparatus

To facilitate measurement of optical transmission through various test fibers, a basic fixture providing (1) means for coupling light into the input end of the fiber under test and (2) means for detection of light emitted from the output end of the fober was designed and constructed. As illustrated schematically in Figure 9, a glass U-tube having two side-outlet arms near the top was provided to contain the test fluids, and the test fiber was looped through the U-tube as shown. The beam from a laser passed through (1) a 90° prism which deflected the beam from a

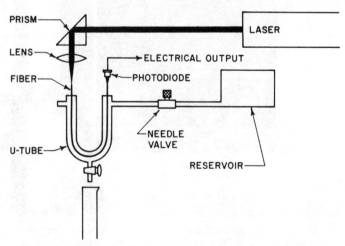

Fig. 9. Schematic diagram of test apparatus.

horizontal path to a vertical path and (2) a microscope objective lens mounted on a three-axis positioner and came to focus on the input end of the fiber. (Note that a laser was used as the light source only as a matter of convenience; non-laser sources could be used equally well). Light emitted from the output end of the fiber was directed onto a PIN (Positive-Intrinsic-Negative) silicon photo-diode. Photocurrent was measured by an electrometer, and output of the electro-meter was recorded on a strip-chart recorder.

A reservoir with a side outlet tube was connected through a needle valve to the fluid input side of the U-tube. Residual test fluids and cleaning solvents were drained from the apparatus at the bottom of the U-tube. The actual test apparatus, which was mounted on a laboratory wall to provide mechanical stability, is illus-trated in Figure 10.

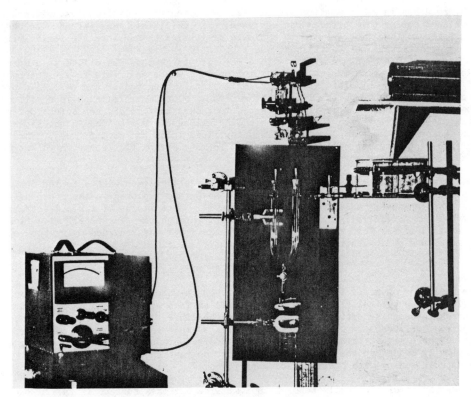

Fig. 10. Optical transmission measurement apparatus.

C. Test Procedure

The fiber to be tested was mounted in the U-tube apparatus and was cleaned with appropriate solvents. Test fluids were prepared by ultrasonically dispersing a known quantity of a hydrocarbon contaminant in a known volume of water. The reservoir was filled with the test fluid to a reference level, and that level was maintained. Adjustment of the needle valve provided regulation of flow through the U-tube at approximately 0.6ml/s. Fluid draining from the U-tube through the side outlet to the left in Figure 10 was collected in a graduated cylinder. The fiber was regenerated prior to introduction of a new test solution by flowing solvents through the apparatus.

D. Sensor Cell Design and Test

A hydrocarbon-in water monitor sensor cell employing the principles established and developed during this program was designed. The cell, which was intended to contain a relatively long fiber coiled to maximize active length and increase sensitivity to low contaminant concentrations, was fabricated in breadboard form as illustrated in Figure 11. A U-shaped groove with a semicircular bypass was milled in a metal plate to provide a common channel for the optical fiber and fluid flow. The ends of the fiber were brought out through small slots at one end of the cell. To prevent fluid loss through the fiber entries, these slots were filled with a silicone compound which had been previously tested to insure that it did not affect optical transmission through the fiber. Female pipe threads were provided in the sides of the sensor cell body to accept fittings for fluid inlet and outlet lines. The top surface of the sensor cell body was lapped to mate with a flat cover which was attached with machine screws and a metal-foil gasket.

Test was accomplished by substituting the sensor cell fitten with a coiled optical fiber treated with reagent "B" in place of the U-tube employed in the previously described apaaratus as illustrated in Figure 12. Test fluids of 1,2,3,4-tetrahydronaphthalene (Tetralin) in water were passed through the sensor cell at a flow rate of 0.6ml/s. Transmission loss through the fiber was recorded using the electrometer and strip-chart recorder as previously described.

DISCUSSION AND RESULTS

The basic component of the reported hydrocarbon-in-water monitor is an organophilically treated optical fiber. An organophilic surface attracts and adsorbs hydrocarbons which decrease optical transmission of the fiber. A surface coating of the type routinely used on column packing materials employed in high-performance reversed-phase liquid chromatography has the desired properties. Such a coating was chemically bonded to the surface of the fiber so that the coating was unaffected by repeated cleaning with powerful solvents (Refs. 4 , 5).

The mechanism of chemically bonding to a sillica or glass surface is fairly well understood, but the precise geometry of the bonding sites is indefinite. In the case of crystalline quartz or certain silicate minerals which have highly ordered crystal structures, the active sites, which are protons covalently bonded to oxygen in the octahedral coordination, are will defined. However, in glass, fused silica, and silica gel, a high degree of crystallinity does not exist, and, in fact, these materials are predominantly amorphous. In such cases, it is not practical to attempt identification of active sites since there is no repetitive structrue.

Figure 13 shows a possible structure of an octadecylsilyl group bonded to a glass surface. It is equally possible that two of the silicon bonds are to hydrogen atoms instead of to the oxygen atoms as shown. It was beyond the scope of the feasibility demonstration to determine the exact bonding condition.

During the experimental program it was observed that chemical treatment of the op-

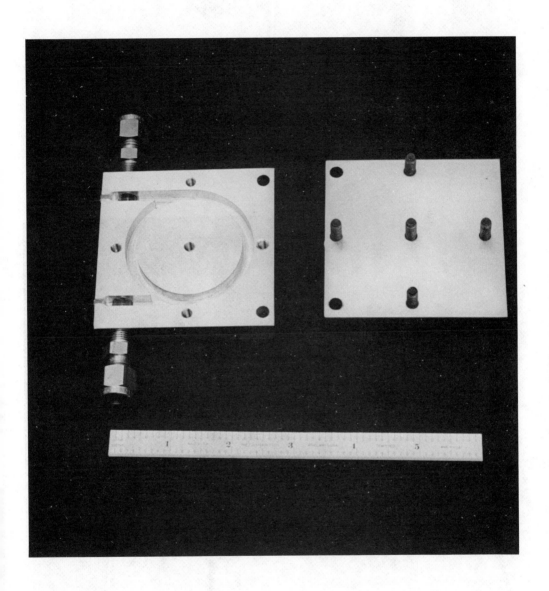

Fig. 11. Sensor cell with cover removed

158

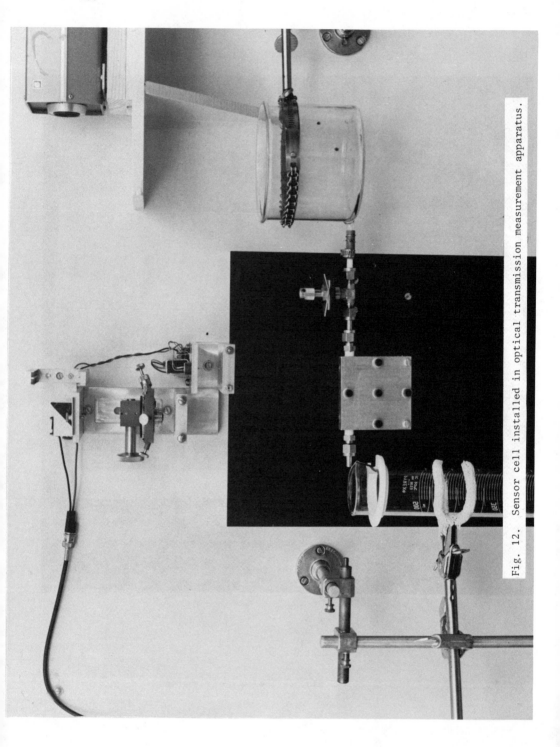

Fig. 12. Sensor cell installed in optical transmission measurement apparatus.

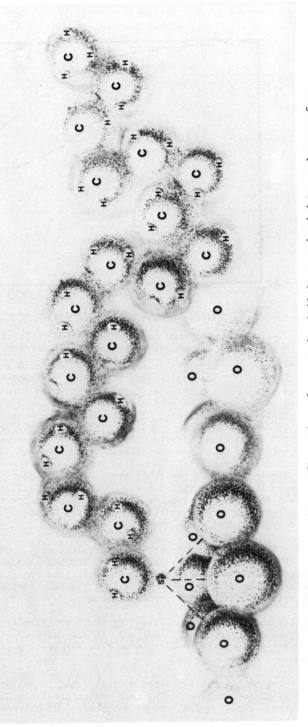

Fig. 13. One possible configuration of an octadecylsilyl group bonded to a glass surface.

tical fiber surface greatly increased sensitivity to hydrocarbons dispersed in water. Additionally, different treatment reagents produced different degrees of sensitivity.

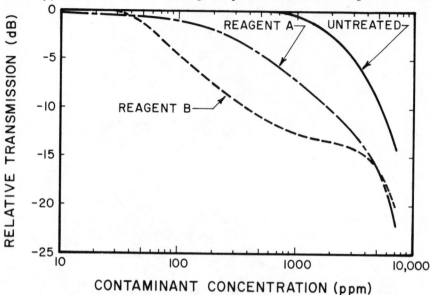

Fig. 14. Optical Transmission Loss Through a 180° Loop of Fiber as a Function of Contaminant concentration and Surface Treatment.

Figure 14 summarizes graphically the data obtained with a single fiber loop in the U-tube apparatus. Transmission loss for a 1000-s exposure to various concentrations of 1,2,3,4-tetrahydronaphthalene (Tetralin) in water flowing at a rate of 0.6ml/s is plotted as a function of contaminant concentration for an untreated fiber and for two differently treated fibers. The untreated fiber shown no response to concentrations of 500 ppm or below. Hovever, the untreated fiber did respond to greater concentrations. A fiber treated with reagent "A" was substantially more sensitive than the untreated fiber and had detectable response to 10-ppm concentrations. The fiber treated with reagent "B" was less sensitive in the very low and very high concentration ranges than the fiber treated with reagent "A", but was much more sensitive in the intermediate range. Observed decreases in optical fiber transmission of 4.5 dB following exposure to 100-ppm hydrocarbon concentrations are indicative of potentially high sensitivity. The data presented in Figure 14 clearly establish feasibility of the method.

The curve of Figure 15 illustrates data obtained with a coiled fiber, treated with reagent "B", installed in the previously described sensor cell. Measurable response to 30-ppm contaminant concentrations and very good sensitivity to concentrations of 100 ppm and above were observes. A slight tendency toward saturation was evident at high values of contaminant concentration.

It was observed that after prolonged exposure to hydrocarbons dispersed in water, a saturation effect occurred. However, in all cases it was possible to restore the fiber to its pre-exposure condition by flowing cleaning solvents through the test apparatus.

Use of an organophilically treated optical fiber to measure hydrocarbons dispersed in water is a novel concept. The fact that appreciable sensitivity to relatively low concentrations of hydrocarbons was observed clearly establishes feasibility of this measurement concept.

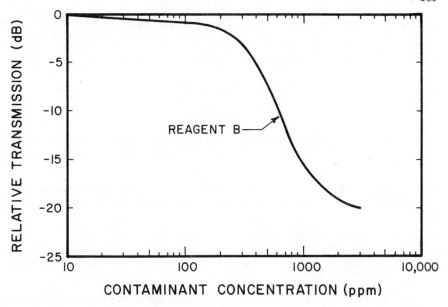

Fig. 15. Optical Transmission Through a Coil of Fiber in the Breadboard Sensor
Cell as a Function of Contaminant Concentration

A hydrocarbon-in-water monitor employing the principles described in this paper
would have a number of significant advantages over existing methods of analysis.
For example, the time-consuming sample preparation steps of solvent extraction,se-
paration from water, drying, and quantitation by ultraviolet or infrared spectrometry would be eliminated. Since an optical fiber hydrocarbon-in-water monitor would
be direct reading, no calculations would be required to obtain a contaminant concentration value, thus indicating that such a monitor could be operated satisfactorily by semi-technical personnel. Moreover, because the sensor cell is of relatively simple construction and because optical fibers can be batch treated in relatively large quantities, a hydrocarbon-in-water analysis instrument employing
the organophilic optical fiber concept should be relatively inexpensive and easy to
maintain. Regenerability of the fiber indicates that the instrument should have
long life.

CONCLUSIONS

It is concluded from the observations made and data acquired during the reported
research program that the concept of measuring hydrocarbons dispersed in water
using organophilic optical fibers is feasible. Organophilic surface treatment significantly increases the capability of an optical fiber to detect hydrocarbon contaminants in low concentrations. To be detected by the organophilic fiber system,
a contaminant must have refractive index greater than that of the optical fiber.
Simple cleaning with commonly available solvents restores the fiber to its original
condition following saturation with the hydrocarbon being measured. In the experimental program, hydrocarbon concentrations as low as 10 ppm were detected, and excellent sensitivity to concentrations of 100 ppm and above was achieved.

ACKNOWLEDGEMENTS

The work discussed in this paper was sponsored by the Environmental Monitoring and
Support Laboratory of the U. S. Environmental Protection Agency, Cincinnati, Ohio,
through its Grant R-804368-01.

Portions of the reported work were conducted in the facilities of the U. S. Army Fuels and Lubricants Research Laboratory located at Southwest Research Institute. This laboratory is operated under contract for the U. S. Army Mobility Equipment Research and Development Command at Ft. Belvoir, Virginia.

REFERENCES

1. Allen Cywin, "Federal Research and Development Program for Oil Spills," Proceedings - Joint Conference on Prevention and Control of Oil Spills, API and FWPCA, 1969, PP. 15-22.

2. H. S. Silvus, Jr. and F. M. Newman, "Development of Oil-In-Water Monitor, " Proposal No. 15-1487, Southwest Research Institute, San Antonio, Texas, 13 October 1975.

3. H. S. Silvus, Hr., F. M. Newman, and G. E. Fodor, "Development of Oil-In-Water Monitor,"Interim Technical Report, Grant R804368-01, Environmental Monitoring and Support Laboratory, U. S. Environmental Protection Agency, Cincinnati, Ohio, 11 November 1977.

4. C. J. Bossart, "Gas Chromatography Column and Method of Making Same,"U. S. Patent 3,514,925, 2 June 1970.

5. J. J. Kirkland, "Microparticles with Bonded Hydrocarbon Phases for High-Performance Reverse-Phase Liquid Chromatography," Chromatographia, Vol. 8, No. 12, Dec. 1975, pp.661-668.

12

HAS THE CLEAN AIR ACT REALLY STOPPED INDUSTRIAL GROWTH?

by

Carleton B. Scott
Union Oil Company of California

On August 7, 1977, the United States acquired a new Clean Air Act when President
Carter signed the 1977 Amendments to the Clean Air Act.

This new Act has great potential for improving the environment. It also has great
potential for disrupting the economy.

Wisely enforced, there can be little doubt that the Act will indeed accelerate
progress toward cleaner air.

Mechanically enforced, without a deep understanding of our Nation's economic
system -- and the economic systems of our competitor nations, there can be little
doubt that the Act will hurt our economy and hurt each of us as individuals.

People in our enforcement agencies have an awesome responsibility. Nearly all of
them, I am sure, will be trying to maximize the benefits and to minimize the risks.
Their problem, however, will be in trying to find out how they are doing.

The benefits should be rather obvious. Each pound of pollutant kept out of the
air certainly can be counted.

But the risks will require care and skill to be seen. Disruptions caused by the
Act are not likely to be obvious or dramatic. Certainly the most serious disruptions will be caused by project delays, and these will come slowly, quietly and
with little fanfare. But as they come, these delays inevitably will trigger three
very dangerous effects:

1. They will weaken domestic industries and make us more dependent upon foreign
   raw materials and foreign products;
2. They will cause more job opportunities to be exported; and
3. They will further weaken the dollar, and make imports increasingly difficult
   to pay for.

Let me focus on two specific parts of the Act where risks will be greatest and
impacts on industry the most profound:

The first is the process that must be followed in revising each state's State
Implementation Plan (SIP) that spells out how the ambient air quality standards
will be attained in that state.

The second is the process for obtaining air permits for any new facility under
the Nonattainment and Prevention of Significant Deterioration (PSD) provisions of
the Act.

SIP Revision Process

Attached as EXHIBIT I is a summary of the process to be followed in revising the States' SIP's.

The essence of this process is twofold: 1) that emissions must be reduced in areas where ambient air standards are not yet attained (nonattainment areas) so as to achieve the standards by December 31, 1982 or -- under special circumstances -- by December 31, 1987 for carbon monoxide and photochemical oxidants; and 2) that emissions in areas where ambient air standards are attained (attainment or PSD areas) must be kept from increasing by very much so that air quality does not "significantly deteriorate" from existing air quality.

There are five key requirements for accomplishing this process, and all revised SIP's must have them:

1. Tighter emission limitations on existing stationary sources in nonattainment areas;
2. PSD Preconstruction Review on all permit applications;
3. New Source Preconstruction Review on all permit applications;
4. Tighter controls on vehicle emissions; and
5. Transportation controls.

Let's talk only about the first three, as these are the most important to industry.

1. Tighter Emission Limitations
    Existing facilities will be the most affected by new, tighter emission limitations.

The Act says that emission regulations must be tightened for all pollutants in all areas where the ambient air quality standards have not yet been attained. New regulations calling for Reasonably Available Control Measures must be required as part of the revised SIP's due into EPA for approval December 31, 1978.

And if reasonable measures will not be adequate to meet the ambient air standards, then progressively tougher measures will be required, some of which can well be unreasonable measures.

The message is clear: Much tougher regulations covering existing operations are on the way, and much more money will be spent on controls.

2. PSD Preconstruction Review
    The Act further says that even if the ambient standard for some pollutant is not exceeded, the amount of that pollutant entering the air should not be allowed to increase by any significant amount. The Act specifies the small allowable increments of deterioration.

In areas near population centers the very small allowable increments of air quality deterioration may well have been used up by earlier projects or by population growth. In such a case, any further industrial growth will be blocked unless the applicant can show that his project will not cause any increase in emissions , or that he can bring about clean-up of somebody else's emissions by an amount that equals or exceeds his potential emissions.

The message here is also clear: PSD review is going to take a lot of time, and cost a lot of money. There is no assurance at all that a project can successfully get through this process in or near most populated or industrialized areas.

3. New Source Preconstruction Review
    While PSD deals with pollutants for which the ambient standard is <u>not exceeded</u>, New Source Review deals with those pollutants for which the standard is <u>exceeded.</u>

The essence of New Source Review is simple: No project shall be allowed if there will be <u>any increase</u> in emissions of any pollutant for which the standard is exceeded.

And if an applicant insists upon building a new facility, or modifying an existing facility in a nonattainment area, then in addition to using emission controls that will provide for Lowest Achievable Emission Rates (LAER), an applicant must also offset any increase in emissions from his project by even greater emission reductions obtained by overcontrolling some other source in the area, or by buying up some source that emits that same air pollutant and shutting it down.

<u>Now here is a most important point</u>: Future industrial growth is directly dependent upon the availability of offsets. And if tighter emission controls are required as part of the revised SIP, as most certainly will be the case in essentially all nonattainment areas, <u>then the only viable offsets that remain will be the shutdown of existing sources</u>.

The implications for future industrial growth, and for future employment trends, are obvious.

But keep in mind the fact that states <u>must</u> develop revised SIP's by December 31, 1978 that incorporate these provisions. If they do not -- or if a state's SIP is considered inadequate to attain ambient air standards by 1982 or 1987 -- then two things happen:
1. Some or all federal grants are cut off to the state or local community; and

2. EPA will simply ban all further construction requiring air permits as of July 1, 1979. And that ban will stay in place until the county, the state or EPA adopts an acceptable SIP that <u>will demonstrate attainment</u> of the ambient air standards by late 1982 or 1987.

EXHIBIT I shows some additonal requirements that a revised SIP must meet, but the important thing to keep in mind is that these requirements we have been talking about -- Tighter Emission Limitations, PSD Review and New Source Review -- are going to be very difficult to meet and to comply with, and even then the risk of a ban on new or modified construction is going to be very real.

Nonattainment and Prevention of Significant Deterioration

At this point the Nonattainment and PSD permitting process summarized in EXHIBIT II becomes effective.

Starting on the left with the initial concept of a new facility or a modification to an existing facility, this roadmap shows the 33 determinations and the 12 studies or actions that the project may have to satisfy, and the 22 ways by which the project may be denied.

Now I don't intend to bore you with all the details of this process. But a few points should be kept in mind:

*   Essentially all projects, regardless of location, must go through this process. The time required looks like a minimum of two years, and there

is a distinct possibility it could be several years.

- This process applies to projects of all sizes because, even in attainment areas, the triggering level of emissions can be as low as 23 pounds per hour of <u>potential</u> emissions -- <u>as if no emission controls were to be used.</u>

- Offsetting emission reductions are the key to this process, certainly in or near most populated areas, yet the only viable offsets that are likely to be available after December 31, 1978 (when the revised SIP's are due), will be offsets brought about by the shutdown of existing facilities.

Combining the key features of the PSD/Nonattainment permitting process with the key features of the SIP revision process, we can see a most perplexing question:

With growth in or near most populated areas being dependent upon the availability of emission offsets; with offsets being severely limited on or shortly after December 31, 1978 when revised SIP's are due for submittal to EPA; and with a ban on permitting in nonattainment areas beginning July 1,1979 if the revised SIP is inadequate; how can industry grow?

Or has the Clean Air Act really stopped industrial growth?

I honestly don't know. I think all I can say is that these requirements are extraordinarily difficult, and that each company will have to decide for itself how these new provisions will affect its ability to grow -- or to stay in business.

We have done this, and here is how we see these provisions affecting our ability to provide energy:

Domestic Crude Oil Production
We see no way to develop a new oilfield or to expand an existing oilfield in most of California. And a similar problem apparently can be expected in some oxidant nonattainment areas east of the Mississippi.

The reason is that after installation of additional controls under the revised SIP's, there apparently would be no viable offsets for the small but finite increases in emissions. And even in those areas where the revised SIP's could assure attainment of the ambient air quality standards on time, there are genuine doubts about the ability of some applicants, particularly the smaller companies, to work their way through the permitting maze in time to justify the effort and costs.

Enhanced Recovery From Older Oilfields
We also have grave reservations about further development of thermally enhanced recovery operations to produce the large quantities of heavy oil that remain in California's Central Valley and, to some extent, in Southern California.

Again, the problem is the same: Even after the most stringent controls have been installed, there apparently will be few, if any, usable offsets.

Refinery Expansions and Modifications
Refinery expansions and modifications appear impossible in California and may well be impossible or severely impaired along the Gulf Coast and East Coast. And these restraints appear to include even modifications:

- To install more octane-generating facilities for lead-free gasoline or for phasedown of lead and manganese in gasoline;

- To enable refineries to handle heavier or higher-sulfur crude oils -- such as that from Alaska's North Slope; and

- To increase production of distillate fuels to compensate for increasing restraints on use of higher-sulfur residual oils.

Expansion of Marine Terminals

And, of course, for every barrel of crude oil not produced or refined domestically, there must be a replacement barrel imported from abroad. But here again the problem is the same.

Expansions of marine terminals to handle imports to cover the inevitable shortfalls of crude oil and refined products are subject to both PSD and Nonattainment provisions, and again we foresee the same problem: After the new SIP provisions are in place, only facility shutdowns can be depended upon for the necessary PSD incremental "room" and Nonattainment offsets. And we doubt that enough such shutdowns can either be found or tolerated in light of current and future energy needs.

I could go on, but all of these examples illustrate the underlying issue:

- Unless the 1977 Amendments to the Clean Air Act are enforced with great care, and perhaps amended by Congress, the United States will inexorably become more dependent upon foreign sources of all raw materials, including crude oil, and even worse, more dependent upon foreign sources of finished and refined products.

- And this means that jobs will be exported even more rapidly than they are now, and that our Nation's ability to find the money to pay for these raw materials and products will become increasingly strained.

Now we all know that there is a finite probability that reason ultimately will prevail, if for no other reason than the fact that the Nation will not tolerate large numbers of people being out of work for very long after July 1, 1979, when the moratorium on new or modified projects goes into effect in nonattainment areas.

But the most serious question appears to me to be: What do companies do in the meantime -- until reason can prevail?

And the answer, of course, is that companies will be able to do very little, and therein lies the danger:

- Projects will be delayed.

- Projects will go into limbo.

- Plants won't be modernized.

- Shortages may well occur.

- And, inevitably, imports will increase.

- And quietly -- bit by bit -- jobs will be exported, and our industrial base will be eroded -- inch by inch.

And ladies and gentlemen, in the very competitive world in which our Nation must operate -- economically as well as militarily -- such a situation simply cannot be permitted.

Each of us -- you, I and all of our many associates, have to turn this situation
around. We must find a way to realize the many benefits of the 1977 Amendments
without risking the very real dangers of those Amendments, the worst of which is
delay.

Let me be so bold as to suggest four specific steps that I think we can take to
help make the new Clean Air Act work:

1. We must insist upon honest ambient air quality standards, honest plans for
    achieving those standards, and honest regulations that will achieve those
    standards.

    Each one of us has a clear personal responsibility to see that county and state
    regulations on emission limitations, as well as on PSD and New Source Review,
    are rational, achievable, and really do something worthwhile for air quality.

    And if those regulations cannot be made productive, or if they are irrational
    or unachievable, then each of us has a similar personal responsibility to let
    our elected officials -- at all levels -- know about our concerns, and about
    our needs.
2. We must develop much better pollution control technology. The public obviously
    wants the environment to be cleaned up, and all of us involved with emission
    sources have to be more effective in developing technology to clean it up.
3. We must stress emissions planning and emissions management in all of our com-
    panies. And we must put emissions planning on a par with the technical plan-
    ning, marketing planning and financial planning that we have been doing for so
    many years. Because it makes no sense at all to do a good job of technical,
    marketing and financial planning for a new project only to have that project
    denied because we could not get the necessary permits.
4. And finally, we must find every possible way to sensitize our legislators to
    the significance of their actions and the very real environmental benefits --
    and risks -- of those actions.

Because in our competitive world, I just don't think there is any other course of
action.

SECTION IV

SOCIAL IMPACT AND SAFETY OF ONSHORE AND OFFSHORE ACTIVITIES

Crude Oil Transportation and Oil Spills : a Brief Overview
of the Implications for Southern California

The LNG Decision Problem

The Management of Offshore Federal Energy Resources in the
United States

The Regulatory Nightmare : Catch 22 !

OCS Impact on Recreation -- the Southern California View

Effects of a Power Plant Effluent on Intertidal Organisms
at Hymboldt Bay, California

13

CRUDE OIL TRANSPORTATION AND OIL SPILLS: A BRIEF OVERVIEW
OF THE IMPLICATIONS FOR SOUTHERN CALIFORNIA

by

K. Kim, L. Hall, and S. Kahane
Environmental Resources Group

Introduction

Southern California waters' exposure to ship traffic is among the highest in the nation. The production and transport of petroleum products account for much of the traffic in these waters, and this proportion is expected to increase in the next quarter century. One of the major concerns associated with the transport of petroleum products is the potential for oil releases into marine waters. Although recent and pending legislation imposes strict standards for tanker construction and operation, some oil spills will continue to occur. Most of them will be small: the majority of spills, on a world-wide basis, have been less than 1,000 gallons. However, a few major spills (over 100,000 gallons) are likely to occur in Southern California as tanker traffic increases. The probability of these large incidents raises concern in relation to North Slope oil importation, offshore drilling, and other related projects proposed for this area.

This paper presents a brief overview of some of the main areas of concern associated with crude oil marine transportation for Southern California. Specifically, the topics chosen are tanker accidents and oil spills, capabilities for responding to spills, and the air quality impacts associated with spill incidents. Tanker ac-- cidents and oil spills are discussed in relation to spill prediction techniques, the relationship between accident location and spillage, and the recent data-base for California waters.

Response capabilities to spill incidents in Southern California waters are evaluated for adverse and calm sea conditions. Recommendations are made for improvement of overall response. Finally, this paper addresses the air quality impacts from spill evaporative hydrocarbon emissions on onshore ambient ozone levels. Because of al- ready poor ambient air quality in Southern California (particularly with respect to ozone levels), large offshore spills could significantly impact critical onshore air areas.

Tanker Accidents and Oil Spills

There are three characteristics common to most oil spills: most spills are small, most of the spillage results from a few large accidents, and the size range of the historical spills is very large (several orders of magnitude). These characteris- tics suggest that a single-number predictor of spillage volume for a particular scenario is not very meaningful. An average spillage predictor may be appropriate if the scenario were spread over many years. For any given year, however, the actual spillage would rarely approximate the average (Beyer and Painter, 1977).

In view of the limitations of such a statistical technique, its wide acceptance is surprising. A substantial portion of the existing literature uses comparisons between modes based in some fashion upon the "average spill volume" statistic (Stewart, 1977). When one sees a statement like "X barrels spilled per Y barrels handled" used to arrive at a statement like "therefore, Z percent of the oil handled by project 'P' will be spilled," one is dealing with the "average statistic". Implicit in such statements are the dual assumptions that spill frequency will be in rough proportion to the volume handled, and that the average volume per spill incident can be determined accurately from the available data (Stewart, 1977). These assumptions are not necessarily supported by the data. Analyses are usually based upon small samples and superficial estimation techniques. The assumptions used are not usually stated explicitly and the validity of the assumptions is not addressed. Further, due to the absence of a recent data-base, many current spill predictions are derived using spill factors from historical data (for the world fleet) that may be close to ten years old.

Becasue of the lack of available data for specific geographic regions, many accident spill impact studies for U.S. waters and U.S. tankers are all still based on statistical probabilities that reflect the 1969-1972 world tanker fleet experience. The U.S. tanker fleet has a much better safety record than the world average. As an illustration, the U.S. tanker fleet during the 1969-1970 period comprised 10 percent of the world tanker fleet and accounted for only seven percent of the total casualties and less than one percent of the total spillage (Keith and Porricelli, 1973).

Tables 1 and 2 present tanker accident and oil spill probabilities based on the world tanker fleet experience, as compiled by Porricelli et al. (1971). It should be emphasized that these probabilities are by no means absolute numbers. They serve only to show relative magnitudes and distributions of spill potentials among various scenarios. Further, the probabilities in Tables 1 and 2 do not account for continuing technological improvements in marine transportation of petroleum products. Beyer and Painter (1977) note that the Western Oil and Gas Association estimates such improvements will reduce future spill incidents by 25 to 50 percent of the historical rates.

Several interesting conclusions can be drawn from the data in Table 1. The principal cause of accidents at sea is structural failure, followed by fires, explosions, and breakdowns. Collisions, rammings and groundings dominate accident causes in and near ports, in coastal waters. Collisions predominate in harbors, and are significant at harbor entrance areas and in coastal waters as well.

Not all vessel accidents result in oil spills (as indicated in Table 2). Further, some accidents are more likely to produce larger spills than others, once a spill occurs. For example, the data for the world tanker fleet show that while collisions account for about 30 percent of all accidents where there is a spill, the amount spilled is only eight percent of the total. Breakdowns and structural failures, on the other hand, released over half the total oil spilled while accounting for only about 20 percent of all spill accidents (Port of Long Beach and Public Utilities Commission, 1976; Porricelli and Keith, 1974).

A similar relationship exists between accident location and amount of spillage. Twenty percent of the accidents at sea result in spills; however, 56 percent of the total oil spillage occurs at sea. On the other hand, while 23 percent of the accidents in coastal waters result in spillage, the volume spilled is only 14 percent of the total oil spilled. Also notable are the relatively small amounts of spillage associated with accidents at piers and in harbors. Approximately 16 percent of the accidents that occur at piers or in harbors results in spillage. These two causes, combined, account for only 10 percent of the total spillage (Port of Long Beach and Public Utilities Commission, 1976; Porricelli and Keith, 1974).

TABLE 1

Tanker Accident Expectancy Per Thousand
Voyages, by Location and Cause

Location

	Piers, Harbors Entrances	In Coastal Waters	At Sea	All
Collisions, Rammings, Groundings	2.34	0.47	0.05	2.86
Structural Failures	0.03	0.01	0.61	0.66
Fires, Explosions, Breakdowns	0.28	0.11	0.45	0.83
Totals	2.65	0.59	1.12	4.36

NOTE: Totals do not add due to rounding.

Source: Port of Long Beach and California Public Utilities Commission, 1976.

TABLE 2

Probability of an Oil Spill After a Tanker Accident
Specific Accident Types and Locations

	Piers, Harbors Entrances	In Coastal Waters	Open Sea	All
Collisions, Rammings, Groundings	0.138	0.379	0.133	0.194
Structural Failures	0.900	0.250	0.201	0.236
Fires, Explosions, Breakdowns	0.247	0.182	0.078	0.147
Totals	0.178	0.341	0.148	0.192

Source: Porricelli et al., 1971; and U.S. Coast Guard, 1972.

More recently, Socio-Economic Systems, Inc. (1977) and Martingale, Inc. (1977) performed an analysis of tanker accidents in California coastal waters and harbor areas. The data-base used was the U.S. Coast Guard Pollution Incident Reporting System (PIRS) records for the 1973-1975 period. The most striking result of the analysis was the significance of fueling operations to overall pollution levels; the three largest spills in the data were all related to bunkering or fueling operations, and fueling contributed well over half the total volume spilled in the 1973-1975 period. A further breakdown of the data reveals that the bunkering accidents were due to either equipment failure or personnel error, the dominant causes of spills during the 1973-1975 period.

Also of interest is the difference in the spill records for Northern and Southern California waters. Table 3 summarizes the tanker traffic and spill data for these areas for the 1973-1975 period. In general, experience in Southern California waters shows an apparent tendency towards smaller spills (despite the fact that the PIRS data show that two of the three largest spills occurred in the area). Southern California experience also displays a slight tendency for the frequency of occurrence of spills to improve with time relative to traffic. Correlating tanker traffic with number of spills indicates that for Southern California there were about 34 spills per 1,000 port calls, and about 8 spills greater than 50 gallons per 1,000 port calls. The corresponding frequencies for the North Area were, respectively, 36 and 14 spills per 1,000 port calls.

While the data highlight equipment failure and personnel error as the primary source of pollution in the region, particularly in conjunction with bunkering operations, it must be emphasized that the period did not include any major catastrophic events such as might be expected from ship loss. Such events are highly significant, although relatively rare, because a single Argo Merchant disaster could produce 150 times the typical total annual pollution level represented in the tanker traffic data for California waters. Tanker traffic in California waters is, at present, relatively low on a world scale (e.g., over 50 percent of the world tanker fleet capacity is used on a single route -- Arabian Gulf to Northwestern Europe). Over the 1973 to 1975 period, only 48 tankers were lost world-wide from a fleet of about 4,500. Hence, the fact that no catastrophic events were reported during that period does not imply that California waters are exceptional with respect to lack of navigational hazards or above average operational procedures. The data simply do not address the potential problem of catastrophic ship losses and the potential for increased tanker traffic levels in Southern California waters.

Oil Spill Response Capabilities

International treaties and regulations on oil spill prevention and clean-up are still in the formative stages: many omissions, ambiguities and contradictions remain to be worked out. On the national level, the U.S. legal framework for oil spill response is basically provided by the Federal Water Pollution Control Act of 1970 (PL 92-500) as amended in 1971, 1972 and 1977. While earlier laws vacillated with respect to liability and penalties, PL 92-500 clearly establishes the liability of the spiller for cleanup costs. Ultimate responsibility to the public for cleanup and environmental protection rests with the federal government: with EPA for spills in inland waters, with the U.S. Coast Guard for incidents in the Great Lakes, coastal waters, ports and harbors, and with the U.S. Geol. Survey for abatement of spills within 500 ft. of a spill source at an offshore drilling facility.

Pursuant to Section 311 of PL 92-500, a National Oil and Hazardous Substances Pollution Contingency Plan, developed in 1973 and since amended, provides for a federal response capability at the scene of a spill. Regional, state and local contingency plans stem from the national plan; in theory, and often in fact, these plans mesh reasonably well at the local level. In accordance with PL 92-500 and later EPA

TABLE 3

Tanker Spill Frequencies in California Waters
1973-1975

	1973	1974	1975
SOUTHERN CALIFORNIA:			
Number of tanker port calls	1902	1728	1538
Total number of spills	68	67	39
Number of spills per 1000 port calls	36	39	25
Number of spills over 50 gallons	19	16	8
Number of spills over 50 gallons per 1000 port calls	9.9	9.4	5.2
NORTHERN CALIFORNIA:			
Number of tanker port calls	1337	1101	801
Total number of spills	39	48	31
Number of spills per 1000 port calls	29	44	38
Number of spills over 50 gallons	14	23	10
Number of spills over 50 gallons per 1000 port calls	11	21	12

Source: Socio-Economic Systems, Inc., 1977.

regulations, all onshore and offshore facilities engaged in production, storage, processing, or distribution of petroleum products must also have their own spill contingency plans and equipment. In addition, the petroleum industry has formed a number of very active cooperatives which provide excellent response capability.

When a spill occurs, it is the spiller who must immediately report and initiate response to the incident; local, state and federal authorities report to the scene as quickly as possible to monitor the response effort. If the spiller cannot be identified, or if his response is inadequate in the judgment of the federal on-scene authority, that authority assumes control of the cleanup. In Southern California coastal waters, the USCG 11th District monitors cleanup activities and has ultimate responsibility for protection and restoration of the environment, with evaluation of biological impact and direction of onshore cleanup provided by the State Department of Fish and Game. USCG maintains containment and cleanup equipment at several Southern California ports such as Santa Barbara, Los Angeles/Long Beach, and San Diego. Local authorities and the petroleum industry (individual installations and the industry cooperatives) also stockpile varying quantities of equipment at strategic locations. In San Pedro Bay, for example, more than 50,000

feet of boom of varying kinds and diameters -- provided by industry, port authorities, and the Coast Guard -- can be deployed quickly.

The first concern is containment by the use of booms. Then, various techniques and
equipment are used to recover as much of the oil as possible; these include skimmer
vacuum devices, and sorbent materials. Threatened shorelines are protected where
feasible by berms and sorbents. Ultimately, the recovered oil is recycled when
possible and final disposal of sludge or oily sand and sorbents is usually to a
landfill. The total response effort is monitored by the Coast Guard in accordance
with existing federal, state and local contingency plans. Despite some overlapping
of authority, the mechanism generally works with fair efficiency. When the spill
is of small to medium size (under 100,000 gallons), and when the incident occurs
in daylight in calm waters -- specifically, where shoreward current is 2 knots or
less, winds are under 10 miles per hour, and waves are not over 4 to 6 feet -- more
than 75 percent of the spill can usually be recovered. Another assumption must
be made: that the response capability is present quite close to the spill scene,
since the oil spreads rapidly and speed of response is critical.

Most spills in Southern California waters have been in harbors, where waters are
protected and response equipment is located close at hand, so that recovery has
generally been good. But, despite stricter standards of design and operation, the
likelihood of spills occurring in larger quantities and under less favorable conditions may increase. We, therefore, have to consider what today's technology, and
Southern California's capability, provide us in the way of defense against spills
that occur on the open seas, in adverse weather conditions, in large quantities,
and at locations where equipment is not immediately available.

This picture is not so encouraging, though its negative aspects can be countered by
advancing technology and common-sense preparedness. At present, technology does
not permit us to predict containment of spills in waves of much more than 6 feet
with currents above 2 knots or winds much over 10 miles per hour. Poor visibility
can also make operations more difficult. Distance to a spill scene may delay response to the point where recovery efforts are almost useless under such conditions
(It must be stressed that technology is continually progressing. Equipment recently developed by a Vancouver, B.C., firm is said to be capable of containing
spills in Sea State Four, characterized as rough, choppy water; this claim has not
been validated, however.) A major spill occurring in Southern California coastal
waters during such a storm as this region experiences at least every four or five
years would be virtually impossible of containment with equipment presently available. Further, the use of dispersants and oil herders, presently employed in
European waters in cases where spills are uncontainable, is prohibited in this
country because of reported environmental damage. Again, technological advances
suggest that this limitation might be re-examined; newly developed compounds may
be less toxic than the oil. Another constraint is the lack of quickly available
storage space for large quantities of recovered oil, since a barge of sufficient
size may take many hours to arrive at a spill scene.

Despite these limits to our capabilities, there are steps that can be presently
taken to minimize damage to marine and coastal environments. Booms can be deployed
downdrift of a spill; state-of-the-art retrieval equipment can be used within the
boom area; threatened shorelines can be protected with berms, sorbents, and booms;
and attempts can be made to repair the leak or tow the damaged vessel to harbor.
Obviously, effective response capabilities must be located at strategic points all
along our coast and not only at large harbors.

Beyond this, a number of recommendations seem worth considering:

1. Continuing effort should be made to provide effective, coordinated contingency

plans for the Southern California coastal area.

2. More stringent liability laws and penalties are required.

3. Additional research is needed on prediction of oil-spill drift to aid in protection of shorelines during open-sea spills.

4. A better detection capability is needed for spotting slicks in times of poor visibility.

5. The oily waste disposal problem deserves a better solution than merely landfilling. Research should continue in an effort to provide easily reusable low-bulk and biodegradeable sorbents, and to reuse oily-water waste and sludge.

6. More consideration should be given to the use of dispersants and oil herders, under strictly specified conditions.

7. A greater local response capability is needed for open-sea spills. Equipment equivalent to that used by the Coast Guard should be locally available.

8. Emergency storage capability for recovered oil should be increased. This must be lightweight and transportable by truck or air.

9. Boom fittings should be standardized so that different brands of the same diameter can be used together.

Air Quality Impacts Associated With Oil Spill Incidents

Because of the existing poor ambient air quality in Southern California air basins, only offshore oil spills could significantly affect onshore air quality. In general, ambient air quality is a function of three factors: emissions, meterology, and topography. Topographical features such as basins, valleys or mountains can hinder air movement and cause localized buildups of pollutants. Meteorological conditions play a role, causing short-term changes in air quality. Long-term air quality changes are due mainly to changes in emissions.

When an oil spill incident occurs, the air quality impacts following the spill and prior to cleanup will be a result of the evaporation of hydrocarbons. The rate of evaporation will be a function of the slick radius, the amount of volatile components in the crude, and sea and meteorological conditions. Experiments by Kreider (1971) have shown that hydrocarbons up to C_{12} (b.p. 216°C) evaporate within the first 24 hours. The fractions between C_{12} and C_{15} (b.p. 270°C) were found to take an additional 20 days for complete evaporation. Open sea tests conducted by Sivadier and Mikolaj (1973) demonstrated that evaporation rates increase with sea surface roughness. They noted ultimate weight losses of 15 percent within the first ½ to 2 hours and 20 percent within the first 2 to 6 hours following the spill. A study by Woodward-Clyde (1976) on the evaporation of North Slope (Alaskan) crude predicted a 10 percent evaporative weight loss within the first hour following the spill, under Southern California meteorological conditions. They predict that about 25 percent of the oil would evaporate within the first 24 hours.

The significance of hydrocarbon emissions is that the reactive portion combines with oxides of nitrogen in the presence of sunlight to form ozone, a powerful oxidant which is the primary harmful constituent of "smog." Emissions of hydrocarbons during the morning hours contribute to higher ozone levels later in the day. The federal hydrocarbon standard of 0.24 ppm for the period 6:00-9:00 a.m.

has thus been promulgated to reduce ozone concentrations particularly in the late afternoon when ozone concentrations are highest. Thus, as the hydrocarbon emissions from an oil spill drift inland and encounter nitrogen oxide emissions, downwind communities will be affected by an increase in ambient ozone concentrations.

If one were to assume the unlikely event of a 165,000 ton ship carrying Alaskan crude oil instantaneously releasing the contents of one of its cargo tanks (about 60,000 barrels) under Southern California meteorological conditions, approximately 936 tons of hydrocarbons might be expected to evaporate within the first hour following the spill. This represents about 61 percent of the hydrocarbons emitted per day in the South Coast Air Basin (SCAPCD, 1976). If the accidental spill occurs 12 miles off Redondo Beach in Santa Monica Bay, during slightly unstable atmospheric conditions, with a windspeed of 4.5 miles per hour, Gaussian dispersion modeling would yield a maximum increase in ambient hourly hydrocarbon levels of 28.6 ppm C at the shoreline. This increase in onshore ambient hydrocarbon levels is a worst-case estimate. The Gaussian model assumes unidirectional and constant winds, causing all the evaporative hydrocarbon spill emissions to blow onshore. The Gaussian model is also limited in that it can only provide best estimates, and not infallible predictions.

Present photochemical models used to predict ambient ozone levels are unsuitable for such a large point source of hydrocarbon emissions. The extremely high concentrations of hydrocarbon emissions associated with oil spills are outside the performance range of these models. However, Figure 1 presents a predicted relationship between morning hydrocarbon levels and afternoon oxidant levels. This relationship was derived by plotting the peak one-hour oxidant level from four different cities vs. the 6:00-9:00 a.m. ambient non-methane hydrocarbon levels for the same day (EPA, 1971). The curve drawn represents an approximate upper limit of possible oxidant concentrations for a given level of morning hydrocarbon concentrations.

The air quality impacts from a spill in Southern California waters could be significant to onshore receptor areas. The typical daytime winds blow onshore and may transport the spill emissions to inland receptor areas. The nighttime drainage winds generally show a reverse pattern of air transport. Figure 2 presents a trajectory computed for the hypothetical emissions resulting from a daytime spill at approximately 8:00 a.m. in Santa Monica Bay. The source of the historical data used in the trajectory analysis was the hourly average wind measurements made by the Los Angeles Air Pollution Control District (LAAPCD) during a three-day oxidant episode period. As shown, the inland areas east of the San Gabriel River Freeway are potential receptors. The greatest impact would probably be experienced here during afternoon hours.

Spill emission impacts on ambient ozone levels would be particularly significant on Southern California air basins (e.g. South Coast Air Basin) which presently exceed state and federal ambient air quality standards. The magnitude of these impacts is not certain. It will be determined by the properties of the crude oil, the size of the spill, spill radius, the distance from shore, the time of the incident (whether the spill occurred when the winds blew inland or out to sea), whether the oxides of nitrogen are of sufficient concentration to react with the hydrocarbons, whether there was adequate sunlight to drive the photochemical reactions, and whether the spill response effort was prompt and adequate.

Conclusion

On the whole, coastal waters have a history of smaller operational spills, as opposed to larger casualty spills, which are more often caused by breakdowns and structural failures on the open sea. However, as the production and transport of

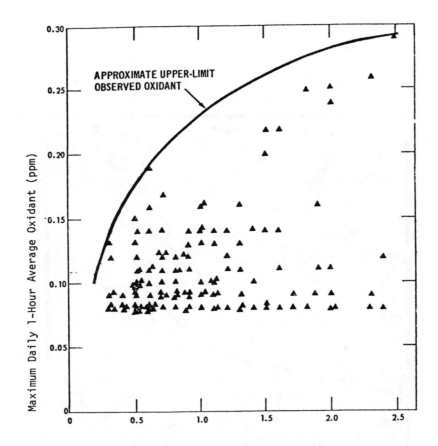

6:00-9:00 a.m. Average Non-Methane Hydrocarbon Concentration
(ppm C)

Source: Environmental Protection Agency, January 1971

Fig. 1. Maximum daily oxidant levels as a function of early morning
non-methane hydrocarbons measured at four different cities.

Δ Peak hourly oxideant concentration per 6:00-9:00 a.m. ambient
non-methane hydrocarbon concentration

180

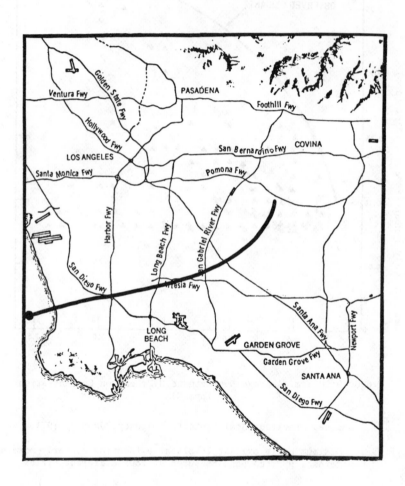

Fig. 2. Air emission trajectory for hypothetical oil spill
in Santa Monica Bay

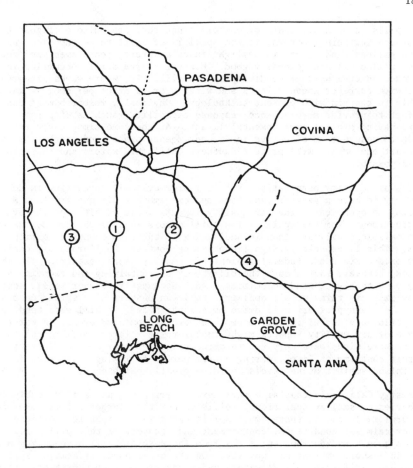

CROSSING FWYS.

① HARBOR FWY.

② LONG BEACH FWY.

③ SAN DIEGO FWY.

④ SANTA ANA FWY.

TRAJECTORY — — — — —

Fig. 3.

petroleum increase tanker traffic in Southern California coastal waters, there may be a greater likelihood of large oil spills in this area.

It is difficult to assess properly the spill potential for California waters, since current spill prediction techniques are based on spill factors derived from historical data on the world tanker fleet; the U.S. fleet has a much better safety record than the world average. In addition, recent tanker traffic and spill data for California waters do not reflect rare catastrophic incidents. Finally, prediction techniques do not take into account future technological improvements in crude oil marine transportation. At present, tanker traffic in California coastal waters is low on a world scale. Correlation of tanker traffic with number of spills suggests that the frequency of spill occurrences in Southern California coastal waters decreases with time relative to traffic.

Large spills in open water are of concern since they are quite difficult to contain. In Southern California waters, if the spill is of small to medium size (under 100,000 gallons) and occurs in daylight in calm waters, containment and recovery of the spilled oil are generally good. However, large spills occurring in open seas under adverse weather conditions -- specifically, where waves are greater than 6 feet with currents above 2 knots and winds over 10 miles per hour -- may be impossible to contain with present technology. Obviously, research must continue in an effort to provide more advance response capability, such as dispersants and oil herders having minimal environmental impact. At the same time, every effort should be made to ensure that effective local contingency plans, and stockpiles of state-of-the-art equipment, will protect Southern California waters and shorelines as much as possible.

From the air quality perspective, spills (large or small) occurring in coastal waters are likely to have a greater impact on onshore ambient air quality than will spills occurring in open seas. The air quality impacts would result from the evaporation of hydrocarbons from the spill. If a spill occurs in daylight hours, the hydrocarbon emissions will be blown inland, and will increase downwind ambient ozone levels. This is especially significant since Southern California air basins presently exceed state and federal ambient air quality ozone standards. It should be noted, however, that though the difficulty of containing and removing a spill increases with higher wind velocities (and consequently rougher seas), some trade-off may exist in terms of air quality: while an onshore wind strong enough to hamper the cleanup effort will drive emissions inland, it will also tend to disperse pollutants and ultimately move them beyond populated areas. A worst-case estimate of air quality impacts might thus involve a large spill in coastal waters while onshore winds were only two to three miles per hour but seas were too choppy to permit containment (i.e., during a calm period following a storm). In such a case, inland ambient ozone levels could be greatly heightened.

In summary, California coastal waters have experienced, and will probably continue to experience, smaller operational spills rather than larger spills resulting from structural failure. If these small spills occur in daylight in sheltered waters or under calm sea conditions, containment and recovery of the spilled oil will probably be good. Hydrocarbon emissions from these spills will have only a minimal impact on onshore ambient air quality. On the other hand, if a major spill occurs in Southern California coastal waters under adverse weather conditions, response efforts may be ineffective. If a catastrophic spill occurs in open seas under such conditions, removal is virtually impossible with present technology. On the basis of world data, the probability of an event such as structural failure in coastal waters is estimated at 1 in 100,000 voyages; the probability of a spill following such an event is 0.25, again for coastal waters. Although such a catastrophe would occur very rarely, subsequent impacts on onshore ambient ozone levels could be quite significant.

References

1. Barker, Charles, General Manager, Southern California Petroleum Contingency Organization, Long Beach, CA. Personal communication March 23, 1977, June 1977 and February 1978.

2. Beyer, A.H., and L.J. Painter. "Estimating the Potential for Future Oil Spills from Tankers, Offshore Development, and Onshore Pipelines." In API: Proceedings 1977 Oil Spill Conference. March 8-10, 1977.

3. Dennis, LCDR Samuel J. "United States Coast Guard High Seas Oil Containment System." In API: Proceedings: 1975 Conference on Oil Prevention Control of

Oil Pollution. March 1975: 365-368.

4. Donahue, Michael. Wm. H. Hutchison and Sons, Wilmington, CA. Personal communication, March 9, 1977.

5. Federal Register, 40(28):6282-3302. National Oil and Hazardous Substances Pollution Contingency Plan (40 CFR 1510), February 10, 1975.

6. Gonzales, John, Supervisor, Crosby and Overton, Inc., Long Beach, CA. Personal communication, March 7, 1977.

7. Herring, Peggy. Clean Seas, Incorporated, Santa Barbara, CA. Personal communication, June 1977.

8. Jaconi, Dennis, Manager. Crowley Environmental Services, Inc., Wilmington, CA. Personal communication, June 1977.

9. Keith, V.F., and J.D. Porricelli. "An Analysis of Oil Outflows Due to Tanker Accidents." In API: Proceedings of Joint Conference on Prevention and Control of Oil Spills. March 1973.

10. Kopeck, Ensign John, Marine Environmental Protection Officer, United States Coast Guard, Long Beach, CA. Personal communication, May and June 1977, and February 1978.

11. Kreider, R.E. "Identification of Oil Leaks and Spills." In API: Proceedings of Joint Conference on Prevention and Control of Oil Spills, June 1971. Cited in: Sivadier, H.O., and Mikolaj, P.G., "Measurement of Evaporation Rates from Oil Slicks on the Open Sea." In API: Proceedings of Joint Conference on the Prevention and Control of Oil Spills. March 1973.

12. Milgram, Jerome H., and Richard A. Griffiths. "Combined Skimmer-Barrier High Seas Oil Recovery System." In API: Proceedings: 1977 Oil Spill Conference. March 1977: 375-379.

13. Porricelli, J.D., and V.F. Keith. "Tankers and the U.S. Energy Situation, An Economic and Environmental Analysis," Marine Technology, October 1974: 340-364.

14. Porricelli, J.D., Keith, V.F., and R.L. Storch. "Tankers and the Ecology." Trans. Soc. Naval Arch. and Marine Engrs. Volume 79, 1971.

15. Port of Long Beach and California Public Utilities Commission. Tanker Traffic Study, Draft Environmental Impact Report, SOHIO West Coast to Mid-Continent Pipeline Project, Volume 3, Part 3, September 1976.

16. Rush, Deputy Watch Commander Raymond. Office of the Port Warden, Port of Los Angeles, CA. Personal communication, June 3, 1977.

17. Sivadier, H.O., and P.G. Mikolaj. "Measurement of Evaporation Rates from Oil Slicks on the Open Sea." In API: Proceedings of Joint Conference on the Prevention and Control of Oil Spills. March 1973.

18. Socio-Economic Systems, Inc. "Crude Oil Transportation and Oil Spills: The Implications for Southern California," Prepared for the Port of Long Beach, July 5, 1977.

184

19. Southern California Air Pollution Control District (SCAPCD). "SCAPAD Emission Inventory," November 5, 1976

20. Standard Administrative Region IX. Coastal Region. Multi-Agency Oil and Hazardous Materials Pollution Contingency Plan, December 1971, and 1974.

21. Stewart, Robert S. "The Tanker/Pipeline Controversy." In API: Proceedings: 1977 Oil Spill Conference. March 1977.

22. Tetra Tech., Inc. Environmental Assessment Report: Crude Oil Transportation Systems: Valdez, Alaska to Long Beach, California. Pasadena, California, April 1977.

23. U.S. Coast Guard Data. Polluting Incidents in and Around U.S. Waters, Calendar Year 1972.

24. U.S. Environmental Protection Agency. "Air Quality Criteria for Nitrogen Oxides, AP-84, January 1971. Cited in: California Air Resources Board, Emissions and Air Quality Assessment, April 1976: 4.54.

25. University of California, Los Angeles. Southern California Outer Continental Shelf Oil Development: Analysis of Key Issues. Prepared by the Department of Environmental Science and Engineering, 1976.

26. Woodward-Clyde Consultants. Air Quality Effects of an Oil Spill in the Port of Long Beach Harbor. San Francisco, California, July 1976.

14

THE LNG DECISION PROBLEM

by

Lloyd L. Philipson
Institute of Safety and Systems Management

SUMMARY

Increasingly large-scale importation of liquefied natural gas (LNG) has been proposed as one of several potential means for mitigating the economic and public
health risks of a major shortage of gas supplies. The wide range of social trade-
offs involved in decisions on LNG importation, whose resolution is now exciting
all levels of government, is described. Of particular concern are public safety
risks imposed at LNG import terminals. The hazards, and the uncertainties that
are associated with them, that give rise to these risks are described, together
with the main features of their modeling. Potential risk mitigating measures, including alternative terminal siting concepts, are then noted. The paper concludes
with a discussion of the problem of public acceptance of the residual risks and a
possible procedure for its resolution.

INTRODUCTION

The public policy debate on the importation of liquefied natural gas(LNG) approaches
the intensity, and exceeds the breadth of issues, prevalent in the debate on nuclear
power. While the central concerns are the availability of a highly desirable fuel
on the one hand, versus the public hazards in its large scale transportation and
storage on the other, numerous other social and economic trade-off factors important to the LNG decision also arise. This paper attempts (1) to outline the main
elements of the most important of these factors, (2) to examine in somewhat more
depth those particular elements that have been most troublesome in the assessment
of the public hazards, and (3) to consider the impact of these latter elements in
the complete decision problem. A more unified, if no doubt idealized, basis for
the LNG decision is thereby portrayed.

LNG, consisting primarily of methane, is produced at a liquefaction plant at such
locations as Algeria, Indonesia, and southern Alaska. Here natural gas is cooled
to a temperature of about -162°C and is thereby reduced to a liquid with about
1/600 of the original volume. In this form it is efficiently carried to import
terminals by specially constructed ships, typically carrying 25,000 cubic meters
of LNG in each of five insulated tanks. (In addition to many foreign import terminals, there are two operating in the United States at the time of writing, and
a third is due to become operational later in 1978.) At the import terminals it
is piped ashore through an insulated pipeline and stored in several large insulated
tanks, each containing, again typically, about 88,000 cubic meters. A vaporization
plant draws liquid from these tanks, heats it to its original gaseous form, and
pumps the gas into a distribution network. The postulated public hazards of LNG
derive from its great concentration of energy (about 2.2×10^7 BTU's per cubic
meter, in quantities of 25,000 cubic meters or more) that might be released in an

accident by a spill and large fire; or by a spill and formation of a potentially large cloud of flammable vapor that might drift at low altitude to a nearby popu- lated area and there ignite. (1)

As sketched in Figure 1, a complete decision analysis for the importation of LNG would include as negative factors the public safety risks deriving from these hazards (taking into account all feasible mitigating measures); but also would in- clude risks to workers, social and economic risks deriving from any potential for unreliability in supply, negative environmental impacts, and, of course, costs and cost impacts. On the positive side would be the economic, environmental and other social benefits of the supply of natural gas provided.

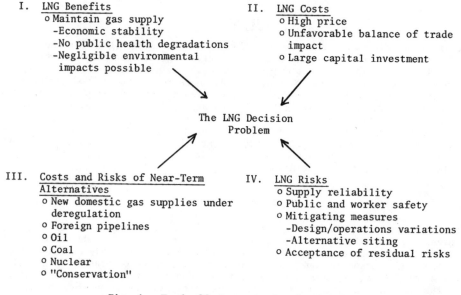

I. LNG Benefits
    o Maintain gas supply
    -Economic stability
    -No public health degradations
    -Negligible environmental
      impacts possible

II. LNG Costs
    o High price
    o Unfavorable balance of trade
      impact
    o Large capital investment

The LNG Decision
Problem

III. Costs and Risks of Near-Term
     Alternatives
     o New domestic gas supplies under
       deregulation
     o Foreign pipelines
     o Oil
     o Coal
     o Nuclear
     o "Conservation"

IV. LNG Risks
    o Supply reliability
    o Public and worker safety
    o Mitigating measures
      -Design/operations variations
      -Alternative siting
    o Acceptance of residual risks

Fig. 1. Tradeoff elements in the LNG decision problem

Alternatives considered would range from no new gas supplies, with impact mitiga- tion through conservation, as feasible; to other means for providing gas or for providing other energy sources capable of replacing gas. These alternatives have specific risks, costs, benefits, and disbenefits of their own. If after rationally balancing all of these factors as well as possible in an overall "quality of life" optimization, some specific level of LNG importation, through specific terminals, and with all feasible safety precautions taken, could be determined to be optimal, there would still remain the question of public acceptability of the residual risks of this importation. Some appropriate decision-making body or bodies must ultimate- ly resolve this issue, under such general and local economic, social and political pressures that then exists.

The subsequent discussion of the LNG decision elements follows Fig. 1. The paper concludes with comments on the acceptability of the principal negative impact of LNG importation, public safety risks, in the context of its benefits. The criti- cal effects of the variance between the assessment and the perception of these risks and benefits are particularly noted, together with some possible means for mitigating these effects.

THE BENEFITS OF LNG IMPORTATION

Natural gas is, of course, a vital source of energy in many areas of the United
States. It has a massive distribution system in place, with a market involving
many industrial facilities,and vast numbers of residential cooking, heating and
air conditioning systems and appliances. It is clean-burning, and, in relation to
the volume and ubiquitousness of its use, safe. Historically, it has also been
cheap, but this is unavoidably changing. An indefinitely large supply of natural
gas exists throughout the world. Domestic supplies have become short, however, at
least in part because of artificially low prices of gas maintained by government
regulation. The American Petroleum Institute states, for instance, that the United
States consumed 63% more domestic gas in 1977 than it added to its reserves, although the addition was the greatest in one year since 1968. (2)

The importation of LNG is one means for compensating for the domestic supply shortage in the near term (others will be discussed later). It can do this with negligible societal and environmental impacts, other than those deriving from its costs
and localized public safety hazards, discussed below. Local environmental problems
can arise, however, from attempts to minimize safety hazards through remote siting
of import terminals, as will also be noted. Important collateral benefits of LNG
importation can derive from the "cold power" that is available for industrial use,
including electric power generation, chemical processing, and food processing.

COSTS OF LNG IMPORTATION

Because of its special processing and transportation costs, LNG is potentially one
of the most expensive sources of gas, per unit of energy delivered. Its price will
probably range from about $3.50 to 5.50 per thousand cubic feet of gas in the near
future. Domestic gas prices, even if unregulated, will probably remain under this
range for many years (at least if sufficient supplementary supplies are available).
The cost of alternative Alaskan and foreign supplies entering the United States by
foreign pipeline will likely approach that of LNG. (4) The high cost of LNG (and
also foreign pipeline gas) may be allocated to all consumers through "rolled-in"
pricing of the mixed total supply of domestic gas and LNG. The impact on each consumer may then become significant, particularly for the lower economic groups, but
it will not be great in absolute terms. Alternatively, the Federal Government may
require "incremental pricing," under which the total cost of the LNG will be borne
primarily by those lower priority industrial users, who would otherwise be allowed
less or no gas. Industrial (and so ultimately, consumer products) costs will rise
under the latter alternative. No doubt, the use of LNG in the United States will
also be greatly diminished if this pricing philosophy is adopted. The consequences
of this depend on the available alternatives to LNG, considered below.

In common with all other foreign energy supplies, LNG also impacts the country's
balance of payments, the "value of the dollar," and so, ultimately, inflation.
LNG's impact is and should remain small compared to oil, but if its importation
reaches a level equivalent to two or three trillion cubic feet of gas per year, the
non-negligible export of perhaps 10 billion dollars per year would result.

Finally, the investment capital required for the transportation and processing
facilities for LNG importation is not easily obtained. The corporations established to develop and operate the facilities are generally ventures of parent
companies (in California, the public utilities) with equity values smaller than
the amount of capital required. Full equity or normal debt financing is therefore
not feasible. Instead "project" debt financing is sought, in which the principal
for the defined project is secured by assured "in-all-events" tariffs to be paid

188

(perhaps, to some extent, pre-paid) by the consumers through increases in the rate
structure, even if some untoward events prevent the establishment or continuation
of the gas supply. (5) This feature has created difficulties in the minds of some
regulators, and may ultimately also prove to be a major problem in the development
of LNG importation. The only known mitigating measure is government loan guaran-
tees, and Congress has indicated no willingness to do this even for pipeline projects. (4)

NEAR-TERM ALTERNATIVES TO LNG

Alternatives to LNG include other gas supplies able to be acquired in, say, the
next twenty years, replacement of gas by electricity generated from other fuels,
and, perhaps, in part, conservation.

The domestic supply of gas may significantly increase, at least for a time, if
Congress' deregulation bill ultimately allows sufficient investment incentives for
gas well development. There is, in fact, an important school of thought that
holds that very large amounts of developable gas exist in the United States, if its
selling price comes to justify its acquisition. (6)

Synthetic or substitute natural gas (SNG) developed from coal may well provide a
supplementary source of gas in perhaps ten years. This can occur if the price of
other gas supplies rises sufficiently to enable expensive SNG to compete.

North Slope Alaskan and Mexican gas supplies have the potential for satisfying a
large part of the country's needs, albeit at much the same high cost as LNG. Moreover, at the time of writing, there is an evident risk that this potential may not
be realized for at least a considerable period of time. The gas would be transported through pipelines in Canada and Mexico. The Canadian pipeline has severe
financing problems, and the proposed price of the Mexican gas has not yet been
accepted by the United States government due to its disparity with that of domestic gas. This problem could be resolved, however, under deregulation of the domestic gas. Note that a pipeline project's financing can be a greater problem than
that of an equally sized LNG project because all of its facilities are fixed, with
their utility tied to one specific source of supply. These facilities lose all
value when this supply is exhausted. Only the liquefaction plant is inflexibly
situated in the LNG system. Its ships can be transferred to runs from other sources,
and its receiving terminals applied to the gas from these sources.

Albeit at high cost, gas users can be transferred to other forms of energy. It has
been estimated, for example, that replacing all gas uses with electricity in southern California would require more than $10 billion, (7) and for the entire United
States, $500 billion. (8) The larger industries are able to switch, and already
are switching from gas to burning oil or coal. Smaller industries, businesses
and residences will only be able to employ centrally generated electricity if gas
becomes unavailable.

However, the use of coal or oil in large amounts without the production of unaccept-
able air pollution is well-known to be extremely difficult in many areas. Scrubber
technology to reduce emissions from coal-fired plants, in particular, is very costly and only marginally effective. Air pollution in an area increases the risk of
illness and death in the exposed public, particularly among the elderly and individuals with bronchial ailments. It also degrades comfort and pleasure factors in
the overall "quality of life".

Additional risks associated with the increased use of coal are the increased ill-
nesses, injuries, and fatalities that will accrue to coal miners, especially in

underground mines; to railroad workers involved in the transportation of the coal; and to the public traversing railroad and highway crossings. Sagan (9) states, for instance, that a total of about 1.3 fatalities per year may on the average be attributed to a 1000-megawatt coal-fired plant. Assuming each kilowatt-hour requires 8500 BTU's of input heat from the coal, a rough calculation indicates that about 30 shiploads of LNG would be able to supply the 8.5×10^{13} BTU's necessary for generating 1000 megawatt-years of electricity from gas. (It is assumed in this that the greater heating efficiency of gas more or less balances the loss of gas in the transportation and processing of the LNG.) Risk estimates for the once-proposed LNG terminals at Los Angeles and Oxnard, California, indicate that, depending on the conservatism of the analysis, 10^{-3} to 10^{-1} fatalities per year* would be expected to accrue at each port in its handling of 565 shipments per year. (1) Thus, each such LNG terminal, more than equivalent in the energy it provides to ten 1,000-megawatt coal-fired plants, would have a fatalities risk one to three orders of magnitude less than that of only one such coal-fired plant. That is, LNG's risk per unit of energy supplied is two orders of magnitude or more lower than that of coal.

For a 1000-megawatt oil-fired plant, Starr et al state that, considering only pollution effects, a risk of fatality per exposed individual per year of about 6×10^{-6} exists. (10) (Accident risks for the public are assessed as negligible.) Assuming a conservatively large exposed population of 10^5 individuals (as in Oxnard) then leads to a total of 0.6 expected fatalities per year. This is about one-half the risk of the coal-fired plant, and so again exceeds that of an equivalent LNG terminal by roughly two orders of magnitude or more.

The coal and oil alternatives to LNG thus suffer from greater total risks than LNG, as measured by expected fatalities per year. However, the sufficiency of this measure in the public view is arguable. As will be discussed later in this paper, this view tends to focus on unlikely catastrophies, rather than on relatively frequent small accidents and continuous health degradations.

The nuclear-electric alternative is considered to be at least one or two orders of magnitude less risky to people than coal-fired plants. (11) Hence, the nuclear and LNG sources of energy are roughly equivalent in this regard. The public views of the two sources are also similar; both are seen as catastrophic threats, however high their average safety and public health levels may be.**

The final short-term alternative available is that of no new energy development (while waiting for solar and other "soft" energy sources), combined with energy conservation. This alternative clearly minimizes public safety risks deriving from the available energy systems, energy costs, and international balance of payments concerns. Its principal risk is that of the economic and, perhaps, social disruption that may result from a severely inadequate supply of energy, and/or govern-

---

* These estimates do not include risks to workers, but due primarily to the relatively small numbers of workers involved with LNG, these would be negligible additions to the risks presented.

** It should be kept in mind that the foregoing comparative numerical assessments consider only the effects of energy generation processes. Fatalities and injuries also arise in accidents involving the users of gas and electricity. Gas user individual risks appear to be 100 times greater than from LNG importation or nuclear power generation hazards. (12) Electricity user individual risks are four times larger still. The total risks for the populations of gas or electricity (or both) users are therefore clearly far greater than for the population exposed to accidents or health degradations from energy generation facilities.

190

ment actions that may be required to enforce the public's adapting to the inadequate supply. This risk can be expected to be of greatest magnitude for the lower economic groups in the country.

RISKS IN LNG IMPORTATION

The risks in the importation of LNG are of two kinds: those to the reliability of supply, and those to the safety of people.

Supply Reliability Risks
The operational reliability of the several components of an LNG system has been assessed in a recent paper by the author.[13] These components include: gas supply source, liquefaction and storage, transportation, storage and vaporization, transmission and distribution (perhaps also with gas storage). In essence, it is argued in the referenced paper that the LNG system has such great redundancy at every point (and this redundancy should increase as LNG is more utilized throughout the world), that supply should be highly reliable.

The potential impacts of any political unreliability of the sources of LNG will decrease as the number of independent sources increases. LNG vessels have generally been very reliable; as more are built, the chartering of a replacement for any one that may have to leave service for a time will become easier. Due to design simplicity and extensive redundancies, a vaporization facility or a receiving, storage and vaporization terminal and associated gas distribution facilities will have a significant outage only if an external event (such as an earthquake) occurs that causes great physical damage. The probability of such an occurrence is extremely low, as will be seen in the discussion of safety, below. Moreover, siting facilities at multiple locations will mitigate the risks deriving from such events.

Safety Risks
Public safety risks from the arrival, unloading and storage of LNG at receiving terminals have been analyzed in a long series of studies over the last five years.[1] Extensive theoretical and experimental work on the phenomenology of LNG spills, fires and vapor clouds has also been conducted. However, major uncertainties still exist in many important areas of LNG risk analysis, and these uncertainties have exacerbated the public controversy on LNG importation. The modeling and data that have been employed, and the issues that have been raised about them are next reviewed. The conclusion that can nevertheless be drawn on the public risks of LNG are then exhibited. The general results of the very little analysis of risks to LNG workers that has so far been accomplished are also noted.

The risks in LNG importation derive from arriving vessel operations, and from operations of the shore transfer, storage, vaporization and transmission systems.

Risks In Vessel Operations

Figure 2 exhibits the principal steps in a vessel spill risk analysis. As applied to an LNG ship, the inputs define the characteristics of the problem that are relevant to modeling the probability of (1) having an accident near a populated area on shore that (2) results in a major spill. The vapor from the spill may (3) either immediately burn, or perhaps develop a vapor cloud that could ultimately drift to a populated area, and ignite and burn. (4) The immediate fire or ignition of the cloud may, depending on various conditions, result in certain numbers of fatalities (injuries and dollars of property loss have not been assessed). (5) The probabilities of all the events leading to these consequences are than combined with the magnitudes of the consequences to provide the estimated public risks, in terms of expected fatalities, or expected fatalities per exposed individual, per year, from the nearby population.

All procedures employed to date conduct steps (3)-(5) in essentially the same manner. The approach to these steps is outlined below, together with the specific models employed in its conduct. Steps (1) and (2), however, have been treated in two distinct ways: by a modified statistical inference procedure, or by kinematic modeling. It is in these steps of an LNG risk analysis that the largest uncertainties and differences among alternative approaches arise.

Statistical Inference

The first method has been employed by the Federal Power Commission (FPC; now the Federal Energy Regulatory Commission, FERC) (14) for the proposed California terminals, based on work of the Oceanographic Institute of Washington. The Woodward-Clyde risk estimates for the Matagorda, Texas LNG facility use a similar technique. (15). Socio-Economic Systems (SES) used a similar procedure in developing tanker accident estimates for the Oxnard LNG Project Environmental Impact Report (EIR), (16) as well as for the Los Angeles SOHIO Crude Oil Terminal EIR and the Southern California Coastal Oil Spill Study.

The method establishes estimates of the probabilities of occurrence of collisions, grounding and rammings (the kinds of accidents that can lead to ship tank ruptures and spills) for a given LNG vessel from vessel accident report data and statistical data on vessel transits in ports and harbors similar to those of interest. It does this by developing them first for a larger class of ships for which reasonable accident statistics exist, such as oil tankers, and then successively modifying them to account for the anticipated differences in LNG ships and their operations. To date, the modifications have been established from judgement and simple modeling applied to specific scenarios for the port of interest.

Similarly, in the risk analysis step (2) involving the development of tank rupture probabilities and spill magnitudes, past general tanker statistics are again modified to reflect the special design characteristics of the LNG ship and the geometry of its transit. Here, naval architectural considerations of hull strength and tank

Inputs

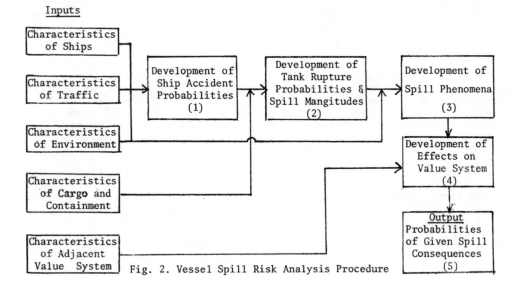

Fig. 2. Vessel Spill Risk Analysis Procedure

location versus penetration forces are involved, as well as the potential striking angles presented by the particular movements of the LNG ship and its tugs as they move through a port scenario.

Kinematic Modeling

The second method for the first two steps of the risk analysis procedure is exemplified by the work of Science Applications, Incorporated (SAI) in its studies (16) in support of the California terminal applications by the Western LNG Terminal Company (WLNG; now Western LNG Terminal Associates); and by A. D. Little Inc. (ADL) in its Point Conception, California, LNG Terminal EIR. (17) SAI and ADL analyze ship collisions by assuming ship motions are essentially random in a zone of interest in a period of time preceding an accident. A kinematic model provides the expected number of collisions per year under this assumption for a port area with specific configurational and traffic characteristics. A calibration to the actual is then made by scaling the model to fit past collision frequencies in these areas.

Rammings and groundings can also be treated in the foregoing models, but are assessed to have either low likelihoods of occurring or of causing tank ruptures, or both, for LNG ships.

The probability of a tank rupture and spill following an accident is developed in the SAI and ADL approaches in submodels of energy exchange and hull rupture dynamics. The former computes, e.g., the expected collision energy lost in vessel motion after impact and the amount remaining to rupture the hull of the struck ship. The hull rupture model then determines whether the cargo containment will fail due to the application of the net energy.

Consequences

As has been stated, all LNG risk assessments conduct steps (3), (4), and (5) in conceptually the same way, albeit with sometimes greatly differing modeling details. A given size and rate of spill leads to specific consequences from an immediate fire or the formation, propagation and ignition of a vapor cloud. Casualties in a specified target value (population and property) structure then are computed. This is done, for example, by describing the value distribution in a grid overlaid on the populated area and then assessing the portions of the grid covered by the range of a given effect, such as thermal radiation above a fatal level. The total value, or a specified fraction of the value in these portions, is then accounted as lost due to the given effect.

The total probability of the sequence of events leading to a specific consequence is then the risk of that consequence. The sum of all the products of all possible consequences (e.g., numbers of fatalities) and their probabilities, is a measure of the overall risk from vessel operations, the expected (or average) number of fatalities from the exposed population, per year. The risk can also be expressed in terms of expected fatalities per exposed individual per year, by dividing by the size of the total exposed population.

When LNG spills on a surface such as the ocean, it spreads into a pool, and begins to vaporize from the heat available at the surface. The primary consequence anticipated from a large LNG spill (say, one full ship tank) is a large fire produced by ignition at the time of the accident of the flammable vapor from the pool. The radius of lethality of the resulting thermal radiation can be a half-mile or more, enduring perhaps a quarter of an hour. "Worst case" modeling of a pool fire is fairly straightforward, with such disagreement as exists focusing primarily on the lethality criterion that should be used. ADL, in the Point Conception EIR, estimates the lethal distance, assumed to correspond to severe burns on the skin, as

7,000 feet from the center of the pool caused by a 25,000-cubic meters (single ship tank) spill onto water. (17) Others have previously estimated 4,800 to 7,300 feet. (3)

The most controversial element in LNG spill consequence modeling is the vapor cloud model, of concern because of the assumed possibility that the vapor might not ignite immediately following an accident. The vapor then forms a cold cloud of gas which is heavier than the warm air around it. The cold gas mixes with air, gaining some heat and cooling some air. While mixing with air the cloud spreads out laterally. If there is a wind blowing, the cloud drifts downwind. The mixing, spreading, and drifting continue until the cloud dissipates completely or until it reaches a source of ignition while it is still flammable, and burns.

Several factors affect the maximum cloud size, how far it drifts, and when it might ignite. These include the spill rate--instantaneous versus the same amount spilling over a period of many minutes to hours, wind speed, and atmospheric stability. The volume of the cloud that may ignite depends on both average and peak gas concentrations. To assess these factors, various theoretical analyses have been made, and experiments have been conducted with LNG spills on land and on water, and with wind tunnel simulations. A major new experimental program is now planned by the Department of Energy. (18) Spills have been both instantaneous and timed-release. Tests have been run at various wind speeds and atmospheric stabilities. These data, along with large amounts of information on plume dispersions from air pollution, chemical warfare, and radioactive fallout studies, provide a data base for LNG vapor cloud models. Experimental data on LNG spills are available so far only for spills of up to 100 cubic meters. For risk assessments, calculations of spill effects start with spills of 1,000 cubic meters and may go up to over 300,000 cubic meters. Experimental data, then, may have to be scaled up by two to three orders of magnitude. Part of the controversy in vapor cloud modeling arises in how to accomplish this.

Methane gas will burn when it is mixed with air, and the gas concentration lies between about five and fifteen percent by volume. In some experiments with LNG vapor clouds, it has been observed that average mixtures of as low as 0.5% gas in air have ignited. This is attributed to fluctuations in the concentration ratio of the mixture. In effect, it implies that although the average concentration may be well below the five percent lower flammable limit, there can be peak concentrations high enough to allow ignition. Some theoreticians have suggested that the peak-to-average concentration ratio can reach a level of fifty. They indicate that an average concentration of 0.1% gas could have an ignitable peak. They then speculate that enough "clumps" of these peaks could exist so that a flame could jump from one to the next, eventually setting off the whole cloud. Most investigators agree, however, that there may be modest peak-to-average differences that could support ignition, but not that it is possible to have a sustained ignition of a whole cloud with 0.1% average concentration. Further, SAI's analysis suggests that, for large LNG spills, the peak concentration is very close to the average. This variation in the peak-to-average ratio used in cloud dispersion models is a major reason for some of the different results of LNG safety studies. One particular, extreme model, for example, predicts downwind cloud travel distances of over 100 miles for the 0.1% concentration level. For a 5% level, however, the same model gives distances of 10 to 20 miles.

Different models have been developed from such different technical assumptions and experimental data. The general formulation for the vapor dispersion process involves several simultaneous partial differential equations for the gas concentration as a function of the spatial and time coordinates. These equations express various physical balance conditions. SAI, at great computational expense, has solved a somewhat simplified set of these equations by numerical integration, for different input factors. All other models assume variations of a Gaussian solution of a highly simplified form of the equation. In all models, parameters are determined by calibration with the available small-scale test data.

No one model has yet been accepted as "best" by the scientific community. In a recent evaluation for the Coast Guard, (19) Havens has compared the structures and implications of the various major models. Distances of travel of a vapor cloud from a 25,000 m^3 spill onto the water under roughly worst-case wind and atmospheric conditions vary from less than a mile to as much as 50 miles. A maximum range of 11-17 miles is most accepted, but, this could be very overstated due to the conditions assumed. Still more recently, the California Energy Commission, basing its judgement in part on the results of some wind tunnel experiments, indicates a maximum range of 3-5 miles (depending on the peak-to-average ratio assumed).[20] In the Point Conception EIR, ADL estimates 4-17 miles (depending on atmospheric conditions). It is understood that in his report of a further appraisal, employing extensive computer sensitivity studies, Havens' view will more closely approach SAI's range of estimates, say 2-3 miles, depending on wind and atmospheric conditions. Overall, it now appears that 3-4 miles may become a generally accepted range of values for vapor dispersion distance from a 25,000 cubic meters spill, an order of magnitude decrease from its earliest predictions.

The probabilities of an accident, a spill and then a pool fire or cloud (it is usually assumed "conservatively" that the less dangerous pool fire will occur in only 90% of the spills)[1] then combine with modeled "credible worst-case" fatalities from the fire or from the cloud, ignited after it travels to a populated area, to produce the risks from LNG vessel operations.

Risks in Terminal Operations

The modeling of fixed facility component failures from internal causes has been accomplished primarily with fault tree models supported by engineering analyses of such specific areas as responses to stresses. The work that has been performed for LNG facilities has been analogous to that conducted in the "Reactor Safety Study"[21] for nuclear plant facilities. The techniques employed are generally accepted for comparative design evaluations but are highly controversial in their ability to provide meaningful absolute risk quantifications. [I]

Statistical modeling combined with engineering analyses has been applied to the responses to external forces by the LNG terminal facilities, including the docked ship, transfer systems, and storage tanks. For example, the predicted probability of occurrence of an earthquake that would establish acceleration forces greater than those designed for is the risk of the spill, and the consequences from it, that would then result. The development of the inferred probability of such an earthquake is not a settled procedure.

Similarly, the effects of the other external events---storms and tsunamis, aircraft, missile fragment and meteorite strikes---have been analyzed in a fashion like that employed for nuclear plants by developing statistical estimates of the probabilities of occurrence of these events and the responses of key facility components of the impacts, the spill magnitudes being lower, for example, for the relatively more likely impacts near the top of a tank.

The modeling for land spills of the consequences of a fire or of a vapor cloud's formation, travel and delayed ignition, proceeds in the same way as for water spills, taking into account an analytical representation of the distribution of population in the area potentially affected. The primary differences arise from

the facts that, first, a major land spill will be confined, in all but the most
extreme cataclysms, by a dike built for the purpose, and, second, the rate of vaporization of the spill, and so fire magnitude or vapor cloud size and potential
travel distance, will be much less than for a spill on water. This is because the
freezing of the ground tends to insulate it so as to diminish the rate of thermal
transfer from it to the LNG. The confinement of the spill by a dike onto a limited
area of ground enhances this property (as does also dike insulating materials,
sometimes provided). In the Point Conception EIR, for example, ADL estimated a
maximum vapor cloud dispersion distance of 1.6 - 9 miles (depending on wind and
atmospheric conditions) for a catastrophic spill of the entire contents of an
87,500-cubic meters storage tank. Here a "low dike" confines a relatively large
impoundment area. "High dikes," as were planned for the Los Angeles and Oxnard
terminals, would limit the vaporization rate much more, and restrict the travel
distance to less than a mile. However, they may be more vulnerable to earthquakes
and airplane crashes destroying them together with the tanks.

Some probabilities may therefore exist for spills to occur whose effects can reach
a nearby population. These probabilities combine with the consequences of these
effects to add to the estimated risks of the facility.

Overall Risk Estimates

The risks arising in the vessel and terminal operations of LNG facilities have
been quantified in several risk assessment studies. These studies have employed
the data and methodologies noted above. Their results for the several facilities,
in terms of the expected number of fatalities per exposed person per year, or,
approximately, the chance of death per exposed person per year, differ by about
a factor of 10 to 1,000. The main cause of the differences is in the calculation
of the probability of a ship accident.

Tables 1 and 2 summarize the order of magnitude overall results, derived after some
interpretations to establish a common basis for comparison, for the three proposed
California terminals. The risks could be thought to be at least marginally acceptable, on the basis of comparison with past "revealed" risk acceptance. (23) Mitigating measures could increase their acceptability further (see below).

Risks to LNG Workers

In a preliminary analysis of the safety of possible offshore siting of LNG terminals,(24) the present author carried out some rough assessments of LNG worker
risks at mainland, as well as offshore facilities. In essence it was found that
individual mainland worker risks (chances of fatality per hour exposure) due to very
unlikely catastrophic accidents should be of the same order as for the public
(greater proximity to the spill and fire being compensated for by survival systems).
Relatively likely minor accidents lead to small risks for the workers (and none
for the public). Intermediate accidents, with their intermediate chances of occur-
ence (e.g., from a light plane crash into an LNG pipeline) and intermediate consequences, create the highest risks for the workers, possibly approaching, in
worst case circumstances, those of the workers in such other hazardous occupations
as steel production, railroad operation and emergency services (roughly 10^{-3} expected fatalities per worker per year). (Offshore worker risks could be an order
of magnitude worse, approaching those of coal miners.) Such risks, though "vol-
untary," are perhaps again only marginally acceptable, implying that all possible
care should be taken in the design and operation of an LNG facility and its safety
systems.

Mitigating Measures

Major accidents with public safety implications are universally accepted to be very unlikely. (A reasonable estimate of the possibility of occurrence of such an accident (from Table 1, say about 10^{-4} per year) implies that it is less than the probability of drawing a royal flush poker hand, if one draw of five cards is made once a year.) However, many public groups and decisionmakers remain concerned that if one should nevertheless occur, major public consequences might result. This extremely conservative view (major consequences require not only the unlikely accident to occur, but also the concommitant unlikely existence of near-worst-case spill rate ignition, wind and atmospheric conditions), has led to a desire for additional risk mitigating measures. These fall into two classes: alternative design and operating plan features, and alternative siting.

Table 1. Order of Magnitude Estimates of Probability per Year of Significant Accident (At Least One Tank Spill; 565 Ship Arrivals Per Year Assumed[a])

Site	SAI	SES	ADL	FPC[d]
Los Angeles	10^{-6}-10^{-3}[b]	---	---	---
Oxnard	10^{-6}	10^{-3}	---	---
Point Conception	10^{-6}	---	10^{-5}[c]	10^{-3}[22]

(a) Present plan for Point Conception involves only 193 ship arrivals. However, this has no significant effect on the order of magnitude estimates given here.
(b) Depends on specific harbor area.
(c) ADL also estimates 5×10^{-5} for a less significant, but still felt to be a public hazard, LNG pipeline accident.
(d) Now the Federal Energy Regulatory Commission (FERC).

Table 2. Order of Magnitude Chance of Fatality per Exposed Person per Year

	Los Angeles		Oxnard			Point Conception		
	SAI	FPC	SAI	FPC	SES	SAI	FPC(a)	ADL
Vessel Operations	10^{-7}	10^{-4}	10^{-8}	10^{-5}	10^{-4}	10^{-9}	10^{-6}	(b)
Terminal Operations	10^{-7}	---	10^{-7}	---	10^{-6}	10^{-7}	---	(b)
Total Chance of Fatality per Exposed Person per Year	10^{-7}	10^{-4}	10^{-7}	10^{-5}	10^{-4}	10^{-7}	10^{-6}	10^{-6}

(a) Now FERC
(b) Not separately given

The first class of mitigating measures has been considered in some detail, part-icularly by SES in the Oxnard EIR.[3] Vessel operator and Coast Guard plans to assure LNG ship safety are extensive.[25] For the shore facilities, the use of storage tanks buried in the ground and an LNG ship-to-shore pipeline buried in the seabed instead of on a trestle can eliminate essentially all risks from earthquakes, airplane crashes, etc. Redundant spill, gas and fire detection systems, and fire fighting capabilities further decrease the already small chances of a limited fire leading to a major one. Finally, thoroughgoing risk management procedures minimize the possibility any design or operation flaw may escape detection and correction.

The belief in the need to eliminate with virtual certainty any potential public consequences through alternative siting remote from significantly populated areas, has, in the past year in California, resulted in the apparent demise of the Los Angeles and Oxnard proposals. A relatively uninhabited, pristine area at Point Conception has now been proposed instead, despite its much greater impacts on en-vironmental factors other than safety, some increase in supply reliability risks, greater time delays, greater cost, and lessened potential for collateral benefits (e.g., application of "cold power"). While individual risks per year are about the same for Point Conception area residents as they are for those near the pro-posed Los Angeles and Oxnard terminals, the total risks from a catastrophe are very much less due to the very small size of the Point Conception population.

However, while Point Conception has received its initial State Government approval, at the time of writing it appears even this remote site may not meet all demands of public interest groups and officials. (Other remote sites had been deemed to be preferred to Point Conception by the California Coastal Commission, each of which would have involved still greater cost and time delays, and new interface problems with other public organizations.) An offshore site continues to be in favor by those groups and officials evidently opposed to any mainland California sites. However, unless a natural island site could be established (and suitable and available areas are very rare off California, at least), the previously ref-erenced study[24] indicates many severe potential problems of public and worker safety would accrue. Other environmental impacts, supply unreliability and even greater cost, financing problems, and time delays would also accrue to an offshore terminal (see the Summary Report noted in Reference 24. These problems have just been also recognized, to some degree, in a newly released study report by the Coastal Commission.[26])

SUMMARY OF IMPLICATIONS OF OBJECTIVE LNG RISK ANALYSES

It thus appears that whereas important LNG accidents are sure to be rare, and pos-sible major consequences extremely unlikely, some public perceptions of the result-ing risks make acceptance of an LNG terminal in a populated area very difficult. Design and operations mitigating measures, though clearly decreasing the chances and often, even more, the possible consequences, of accidents creating these risks, appear to have little effect on these perceptions. Alternative remote sites, in-cluding most offshore possibilities, may only increase all or part of the problem (at least in areas with limited places for such sites, as in California). Is the LNG importation problem therefore unsolvable without significant public disappro-bation?

APPROACHES TO ANALYSIS AND MITIGATION OF PUBLIC RISK PERCEPTIONS

One method of solution of the LNG siting problem is, of course, establishment of the site by fiat by an agency of the government, with sufficiently clear authority that the litigation that would then no doubt arise would be able to be cut short.

The Federal Energy Regulatory Administration (FERC; formerly the FPC) has authority at least approaching this. (FERC still favors the Oxnard site in California, (22) and, conceivably, might yet dictate it if the problem is not otherwise satisfactorily resolved.) Congress could strengthen FERC's capability still further. However, difficulties form demonstrators and even saboteurs fighting what they deem an unpopular decision could still arise.

The alternative is to assess and effect changes in public (and representative public officials') attitudes about the use of a desired site. It is clear that the presentation in some appropriate form of the results of objective risk analyses is insufficient for this. Nor, in general, is the presentation of the various risks of the alternatives, such as have been outlined earlier in this paper. Someone else's prediction of the future, no matter how thoroughgoing, is not often accepted by an individual unless it is closely congruent with his own views, or, rather, feelings, about not only what the future will be like, but often also, what it should be like. Present reality may have little impact on these feelings. Thus, the feelings have to be changed, if true acceptance is to be effected.

Slovic et al of Decision Research (e.g., in References 27 and 28) have assessed experimentally such attitudes as have been described. It may be interpreted that they have found that these attitudes are often insensitive to facts, both those that are presented and those that are omitted. The focus is usually instead on some simple expectation (such as a catastrophic hazard), virtually independent of its likelihood of occurrence.

The fundamental potential means for overcoming the problem is the same as the means by which individuals generally become amenable to the acceptance of higher risks. Increase the perceived benefits to the individuals who perceive themselves to be at risk, to a level where they feel acceptance of the risks is justified.* This has been shown to be an elemental aspect of both "revealed" risks acceptance assessed from historical data, (23) and the acceptance of "perceived" risks assessed in the psychometric experiments at Decision Research. (28)

O'Hare has posed and tested one procedure for accomplishing this.(29) In outline, he proposes an "auction" be held among the populations (or their representatives) at several alternative sites considered for a hazardous facility, who would perceive themselves to be placed at risk. After a maximum possible education on the risks and benefits of the planned facility, each population would be asked to establish by voting or other means the amount of money each individual (or family, or other group) would demand to accept the facility at his location. The bid permitting the lowest total facility cost would "win" the facility; if all bids were so high as to make the facility uneconomic, its dropping would be indicated (much more quickly and economically than has often been the case).

No doubt other procedures are conceivable for accomplishing the same end of determining the incremental benefits required for acceptance of a hazardous facility, such as an LNG terminal. The conclusion of the present paper is that the resolution of the highly complex LNG decision problem may well come down finally to exactly this simple point. Perhaps only in this way will it be possible to establish a true optimization of the overall quality of life where it may be impacted by LNG, or by its absence.

---

* This has long been done implicitly, of course, through decreased local taxation of residents due to the presence of a large, and perhaps hazardous, plant in the community. Explicit incremental benefits are under consideration here, that result in risk-taking individuals coming to believe that they would then be receiving fair treatment relative to others,

REFERENCES

1. Philipson, L. L., "The Systems Approach to the Safety of Liquefied Natural Gas Import Terminals," prepared for the California Energy Resources Conservation and Development Commission, May 1977.

2. Los Angeles Times, "Oil and Natural Gas Reserves Fall Again, Trade Groups Find," 10 April]978.

3. Soci-Economic Systems, Incorporated, "Environmental Impact Report for the Proposed Oxnard LNG Facilities," Draft, August 1976; Final, July 1977.

4. Los Angeles Times, "Uncle Sam May Have to Shoulder Alaska Gas Pipeline," 28 May 1978.

5. Brooker, T. K., "Financing LNG: Who Pays and Who Profits," testimony at the Liquefied Natural Gas Hearings of the Assembly Subcommittee on Energy, Los Angeles, 22, 23, 28, 29 July 1976.

6. Hefner, R. A., III, "Deep Zones--The Andarko Basin," testimony at the Gas Supply Hearings of the Assembly Subcommittee on Energy, Los Angeles, 18, 19 October 1976.

7. Los Angeles Times, "Quick OK Urged for Imports of Algeria Natural Gas," 19 April 1978.

8. Lawrence, G. H., "The Federal Triangle, An Energy Soap Opera of Tragic Scope," Proceedings of the American Gas Association Transmission Conference, Montreal, 8 May 1978.

9. Sagan, L. A., "Health Costs Associated with the Mining, Transport and Combustion of Coal in the Steam-Electric Nature," Nature, v. 250, July 1977.

10. Starr, C., et al, "A Comparison of Public Health Risks: Nuclear vs Oil-Fired Power Plants," Nuclear News, October 1972.

11. Bethe, H. A., "The Necessity of Fission Power," Scientific American, v. 239, no. 1, January 1976.

12. Cohen, B. L., "Weighing the Risks of the Life Today, Los Angeles Times, 4 June 1978.

13, Philipson, L. L., "The Operational Reliability of LNG Systems," Proceedings of the Annual Reliability and Maintainability Symposium, Los Angeles, January 1978.

14. Federal Power Commission, "Pacific-Indonesia Project, Draft and Final Environmental Impact Statements," May 1976 and December 1976.

15. Federal Power Commission, "Matagorda Bay Project, Draft Environmental Impact Statement," July 1977 (incorporates Woodward-Clyde analysis).

16. Science Applications, Incorporated, "LNG Terminal Risk Assessment Studies for Los Angeles, Oxnard and Point Conception, California, and for Nikiski, Alaska," prepared for Western LNG Terminal Company, December 1975-January 1976.

17. A. L. Little, Incorporated, "Draft Environmental Impact Report for Proposed Point Conception LNG Project," prepared for California Public Utilities Commission, 28 February 1978.

18. U.S. Department of Energy, Assistant Secretary for Environment, Division of Environmental Control Technology, "An Approach to Liquefied Natural Gas (LNG) Safety and Environmental Control Research," DOE/EV-0002, February 1978.

19. Havens, J. A., "Predictability of LNG Vapor Dispersion from Catastrophic Spills Onto Water: An Assessment," prepared for the U.S. Coast Guard, CG-M-09-77, April 1977.

20. California Energy Resources Conservation and Development Commission, "LNG Study: An Assessment of Risks," Sacramento, August 1977.

21. Nuclear Regulatory Commission, "Reactor Safety Study (WASH-1400)," NUREG-75/014, October 1975.

22. Federal Energy Regulatory Administration, "Draft Environmental Impact Statement, Western LNG Project," FERC/EIS-0002, April 1978 (Draft).

23.. Starr, C., "Social Benefit Versus Technological Risk," Science, v. 165 (1232-1238), 1969.

24. Philipson, L. L. and G. Donaldson, "Safety of Offshore LNG Terminals: A Preliminary Analysis," prepared for the U.S. Department of Energy, Region IX and San Francisco Operations, August 1977 (Draft). Summarized in "Siting an LNG Receiving Terminal Offshore: Issues to be Considered," by Region IX and San Francisco Operations Offices, U.S. Department of Energy, 10 March 1978.)

25. U.S. Coast Guard, "Liquefied Natural Gas, Views and Practices, Policy and Safety," CG-478, February 1976.

26. California Coastal Commission, "Offshore LNG Terminal Study," Staff Draft for Public Comment, 14 July 1978.

27. Slovic, Paul and Baruch Fischoff, "How Safe is Safe Enough? Determinants of Perceived and Acceptable Risk," Decision Research, Eugene, Oregon, 1978. (To appear as chapter in The Management of Nuclear Wastes, by L. Gould and C. A. Walker, Yale University Press.)

28. Slovic, Paul, "Perception and Acceptability of Risk from Nuclear and Alternative Energy Sources," Decision Research, Eugene, Oregon, March 1977.

29. O'Hare, Michael, "'Not on My Block You Don't' Facility Siting and the Strategic Importance of Compensation," Public Policy, v. 25, no. 4 (Fall 1977).

15

THE MANAGEMENT OF OFFSHORE FEDERAL ENERGY
RESOURCES IN THE UNITED STATES

by

Robert B. Krueger
Law Offices of Nossaman, Krueger & Marsh

Introduction

More than one-third of the approximately 3,500,000 square miles of land in the
United States is owned by the federal government. Most of those federal lands are
located in the western United States, particularly Alaska, where more than 95% of
the land is federally owned. In addition, the United States has jurisdiction and
control over as much as 1.3 million square miles of continental shelf and slope.
Beneath those lands exists an enormous quantity of energy resources. This article
will discuss the procedures by which the federal government manages the exploration
and development of offshore energy resources and the various policy considerations
which underly those procedures.

Federal Offshore Energy Resources

Introduction

The extent of proven offshore petroleum reserves in the world is considerable and
the potential for development is enormous. There are today approximately 30 nations
which have established offshore oil and gas production with aggregate reserves of
approximately 90 billion barrels or over 20% of the world's total reserve figures.
On a worldwide basis, current offshore production is over 6.5 million barrels per
day or between 15 and 20% of the world's total.[1] The Department of the Interior
has estimated that, by 1980, approximately 30% of the oil requirements[2] and 40% of
the gas requirements in the United States will come from our offshore.[2]

Looking at the United States alone, the presently proven reserves of oil and gas
on the outer continental shelf are approximately 5 billion barrels of oil and 35
trillion cubic feet of gas with prospective reserves of an additional 3 to 19
billion barrels of oil and 27 to 97 trillion cubic feet of gas.[3] These figures
do not include state offshore lands, which have to date produced something in
excess of 750 million barrels of oil, and areas, such as Prudhoe Bay, which have
immense reserves.[4]

Disputes Relating to Control over the Outer Continental Shelf

It appears quite clear today that the petroleum resources of the offshore extend
into the continental slope (approximately between depths of 200 and 2500 meters)
and possible into the continental rise (approximately between depths of 2500 and
5000 meters).[5] It is, moreover, foreseeable that within the immediate future technology will permit the development of such resources and even mineral resources
in areas far beyond. This factor and the immense potentiality of the continental
shelves and slopes of the world for other minerals[6] have created the heightened
interest in the location of offshore boundaries both between the United States
and the international community and, within the United States, between the States
and the federal government.

International Disputes

Prior to 1945 there was no internationally recognized appropriation to submarine areas outside of a nation's territorial sea, whether the areas were continental shelf or otherwise. There was a great deal of interest, particularly in the United States, regarding the offshore development of oil and gas, but it was directed largely to lands underlying the territorial sea, the three-mile coastal belt. In 1945, however, President Truman issued his landmark proclamation in which he expressed the view that "the exercise of jurisdiction over the natural resources of the subsoil and sea bed of the continental shelf by the contiguous nation is reasonable and just" and proclaimed:

> "...the government of the United States regards the natural resources of the subsoil and seabed of the continental shelf beneath the high seas but contiguous to the coasts of the United States as appertaining to the United States (and) subject to its jurisdiction and control."[7]

At the same time President Truman issued Executive Order 9633 which ordered that "the natural resources of the continental shelf...contiguous to the coasts of the United States...(be) placed under the jurisdiction and control of the Secretary of the Interior for administrative purposes, pending the enactment of legislation in regard thereto."[8]

It is clear that the term "continental shelf" as used in the Truman Proclamation was intended to be interpreted in its geologic sense.[9] In this sense the continental shelf consists of the natural prolongation of the coastal plain of the continental land mass at least to the point at which it becomes continental slope and drops sharply off to the abyssal plains.[10] It thus appears clear that the Proclamation was intended to cover areas such as the "continental borderland" off Southern California which at points lie much deeper than 200 meters but which are ecologically identifiable as a border of the continental land mass.[11]

The concept of the Truman Proclamation later became embodied in the 1958 Geneva Convention on the Continental Shelf and provided the basis for the doctrine of the continental shelf in customary international law enunciated by the World Court in the North Sea Continental Shelf Cases in 1969.[12]

Federal-State Disputes

Prior to 1947 it was thought that California and the other coastal states owned the land underlying the territorial sea. In California, Texas and Louisiana there had been substantial offshore oil production established under state leases predicated upon this belief.[13] In 1947, however, the U.S. Supreme Court determined in United States v. California[14] that the federal government had "paramount rights in (and) full dominion over the resources of the soil under that water area, including oil.[15] The same principle was confirmed as to other coastal states in the succeeding decisions[16] which brought about the political pressure that resulted in the Submerged Lands Act of 1953.[17] That Act in effect reversed United States v. California by vesting in the coastal states the ownership of lands "beneath navigable waters within (their respective) boundaries,[18] which were defined as lands lying within three geographical miles of the "coast line."[19] It also permitted historic boundaries in the Gulf of Mexico to the extent of three marine leagues (9 miles), which were subsequently established in the case of Texas and Florida.[20]

The Provisions and Policies of the Outer Continental Shelf Lands Act

The Outer Continental Shelf Lands Act[21] was adopted in 1953 as a companion measure to the Submerged Lands Act.[22] It was the first federal act authorizing the leasing of offshore lands and created a comprehensive system dealing with all such lands which might be claimed by the United States. Section 3 of the Act states in part:

> "It is declared to be the policy of the United States
> that the subsoil and seabed of the outer Continental
> Shelf appertain to the United States and are subject
> to its jurisdiction, control, and power of disposition
> as provided in this (Act)."[23]

Congress was aware that "continental shelf" as used in its geologic sense extended only to lands lying interior of the geologic slope,[24] but the Act was not restricted to those lands. The term "outer Continental Shelf" was defined in Section 2 as including "all submerged lands lying seaward and outside of the area of lands beneath navigable waters (title to which was confirmed unto the coastal states by the Submerged Lands Act) and of which the subsoil and seabed appertain to the United States and are subject to its jurisdiction and control." It is clear, therefore, that the Act applies to all lands properly claimed by the United States under international law whether as continental shelf, continental slope or otherwise.[26] For this reason the Act itself does not constitute an assertion of jurisdiction by the United States as to any particular offshore area. It is best viewed as a legislative implementation of the 1945 Truman Proclamation[27] and the 1958 Geneva Convention on the Continental Shelf. It would serve the same function for any future regime for the U.S. offshore such as that proposed for the continental shelf and exclusive economic zone in the U.N. Law of the Sea Conference now under way.

Selection of Lands for Lease

The Outer Continental Shelf Lands Act vests in the Secretary of the Interior the authority to grant mineral leases covering areas of the outer continental shelf not already under lease or withdrawn from leasing under the provisions of the Act.[28] No priorities or guidelines by which the Secretary is to select areas for lease are set forth in the Act and historically the regulations of the Secretary simply authorized the issuance of mineral leases in any given area upon motion of the Department or the request of an interested party after requisite competitive bidding.[29] Further, the regulations, while requiring persons obtaining permits for offshore exploratory work to disclose geological data[30] did not require the disclosure of geophysical data such as the "seis line" obtained by reflection seismology. Because geophysical data is the most commonly used and reliable means of delineating potential petroleum and sulphur prospects in the offshore at the pre-leasing stage[31] the result has been that industry has had a better knowledge than the federal government of the potentiality of offshore mineral prospects. As to petroleum and sulphur, a system has evolved whereby the Department of the Interior and industry cooperated in the selection of areas to be leased, with industry conducting all requisite pre-bidding exploratory work and, upon request, nominating areas of interest to the Department's Bureau of Land Management ("BLM"). BLM has, then, with the assistance of U.S. Geological Survey ("U.S.G.S.") and on the basis of their relatively restricted knowledge regarding the properties involved selected the tracts offered for lease sale.[32]

Since the 1969 Santa Barbara Channel oil spill,[33] which had a massive and very negative impact upon U.S. offshore oil development, the Secretary of the Interior has promulgated regulations requiring substantially greater consideration in the final selection of the tracts for leasing of "the potential effect...on the total environment...and other resources in the entire area" during all phases.[34] The regulations also require the development of special leasing conditions where

necessary to protect the environment and other resources.[35] While there is still
no federal requirement for mandatory public hearings in the coastal states and
cities affected, as there is under the California system,[36] the Secretary has established a procedure of consulting with both state and local governments at the
time of tract selection.[37] In addition, he has established a National Advisory
Board on OCS Policy with representation from these areas.

S.9, which passed the Senate in July 1977,[38] would drastically change the system of
selection of lands for lease, if enacted by Congress. It would give the Secretary
of the Interior broad powers to acquire virtually any form of geological or geophysical information by virtually any means, including in some cases exploratory
drilling, to assist in the evaluation of any outer continental shelf lands.[39] It
would authorize the President to conduct an inventory of all U.S. oil and gas reserves in both public and private lands with the power to conduct both geological
and geophysical operations for this purpose.[40] S.9 would further require that the
Secretary be given access to all geological and geophysical information obtained by
lessees and permittees in the course of exploration, development, or production
under the condition that its confidentiality will be maintained unless the lessee
or permittee agrees to its disclosure.[41]

S.9 also requires that the Secretary review offshore nominations with state and
local governments "which may be impacted by the proposed leasing" and would require
coordination of the program with coastal management programs developed by the coastal
states under the Coastal Zone Management Act of 1972[42] an act to fund and coordinate
coastal planning.[43] It, moreover, would require the Secretary to accept "recommendations...regarding the size, timing, or location of a proposed lease sale or
with respect to a proposed development and production plan" from the governor of
"any affected coastal State" if he determines that "they provide a reasonable balance
between the national interest and the well-being of the citizens of the affected
State".[44] S.9 would require the Secretary to comprehensively plan and schedule
offshore exploration, development and productions based upon a number of considerations, including the "laws, goals, and policies of affected States."[45]

Some of these provisions are salutory, such as those which would give the Secretary
of the Interior a greater informational base at the time of leasing. A number of
the other provisions would, as did the 1974 Deepwater Port Act,[46] extend the influence of the coastal states far beyond their offshore lands and directly affect
the management by the federal government of its resources. Perhaps the most significant factor is that the tiers of study, planning and mandatory determinations
required by S.9 in addition to those imposed by the National Environmental Policy
Act[47] will predictably lead to substantially more delays in offshore development.

A parallel bill, H.R. 1614, was passed by the House of Representatives on February
2, 1978.[48] Although similar to the Senate bill in many respects, it does contain
some differences.[49] One such difference is that the House bill, unlike the Senate
version,[50] does not authorize the Secretary to directly contract for exploratory
drilling by qualified applicants in areas believed to contain "significant hydrocarbon accumulations."[50] In addition, H.R. 1614 does not contain any provision
comparable to the one in S.9 authorizing "dual leasing"; the S.9 provisions would
require separate leases for exploration and development in Alaskan sales.[51]

The Outer Continental Shelf Lands Act requires that an oil and gas lease not contain more than 5,760 acres (9 square miles),[52] but does not contain any restrictions
upon the number of tracts which may be simultaneously offered or the frequency with
which they are offered.

The clear preference on the part of the major oil companies has been for large
offerings of blocks of the maximum size and this normally has been the practice of
the Interior to the present time. Through 1975, for example, the typical sale has

resulted in the leasing of 71 tracts, with each tract covering an average of 4,645 acres.[53] At the present time, Interior is preparing environmental impact statements for the possible sale in June of 1979 of 217 tracts in the Southern California offshore covering over 1,140,000 acres.[54]

With due regard to this pattern of lease offering, the bonus bidding system which has historically been used under the Outer Continental Shelf Lands Act has unquestionably placed a growing amount of stress on the industry,[55] which has been reflected in a rapid growth of the ratio of indebtedness to equity among major oil companies. This stress will predictably grow greater with the ongoing effort to vastly expand outer continental shelf petroleum production. This fact could result in the Federal Government receiving less for the sale of the resource than it would receive if the sales were smaller and spaced further apart. It would also tend to discourage the small company, but it must be borne in mind that the economic considerations pertaining to offshore oil and gas development functionally must discourage the small company;[56] the only way that they can be otherwise is through some form of subsidy. On the other hand, bonus bidding is demonstrably the most simple means of lease disposal and does attract the company that is most economically motivated toward early development.[57]

S.9 would retain the acreage restriction on oil and gas leases, unless the Secretary finds that "a larger area is necessary to comprise a reasonable economic production unit."[58] H.R. 1614 contains an identical provision.[59]

Allocation of Lands - Lease Sales
The Outer Continental Shelf Lands Act requires that oil and gas leases be issued by competitive bidding and authorizes the Secretary of the Interior to call for bidding on the basis of cash bonus with a fixed royalty of not less than 12½% or on the basis of royalty bid with a 12½% minimum and a fixed cash bonus.[60] The practice of the Secretary has been to issue leases with a 16-2/3% royalty and on the basis of the highest cash bonus bid except in a few instances in which a higher flat royalty was stated and there was experimental royalty bidding.[61]

The present language of the Outer Continental Shelf Lands Act is probably sufficiently broad to prevent experimentation with a number of alternate forms of offerings, including bidding on the basis of flat royalty, sliding scale royalty and even net profits.[62] Under all of the administrations since passage of the Act, however, the Secretary of Interior has followed the bonus bidding precedent which does have a number of advantages.[63] It is simple to administer, both at the times of lease allocation and subsequently, it provides an incentive for early development and the flat royalty minimizes the problem of premature abandonment that any surcharge on production poses. It has, however, recently been under a great deal of attack in the Congress, seemingly on the ground that it is favored by the petroleum industry and not used in other major petroleum producing countries where large revenues are received.

A possible alternative to the bonus system which has been used in California is one in which the administering agency would establish a sliding scale royalty and award the lease to the bidder who submitted the highest factor to be multiplied by that scale. Another alternative which was used in the huge East Long Beach (California) Field would be to have bidders compete on the basis of the precentage of net profits derived from production they are willing to pay.[64] Still another possibility is the system used in the United Kingdom[65] and Norway[66] where decisions regarding allocation of oil and gas leases are based not on the size of the direct payments that applicants are willing to make, but rather upon consideration by the government of such factors as the experience and competence of the applicants and the quality of their proposed work programs.

No reliable studies have been done comparing the above systems, however, and whether any of the alternatives to the current system would result in a larger after-tax return to the United States in the development of the Outer Continental Shelf is purely a matter of conjecture. Notwithstanding that, congress has asked that these and other systems be examined and used. Included in S.9 and H.R. 1614 are provisions authorizing the Secretary to conduct competitive bidding on the basis of flat royalty with cash bonus, variable royalty bid with a fixed cash bonus, cash bonus with a diminishing or sliding royalty, cash bonus bid with a fixed share of net profits, fixed cash bonus with a net profit bid variable, cash bonus with a flat royalty and a net profits interest, a work commitment bid based on a dollar amount for exploration with a fixed case bonus, or a fixed royalty or net profits interest or a combination thereof.[67] In addition, the House bill authorizes the Secretary to devise any modification of these systems which he "determines to be useful."[68] One of the major differences between the two bills, however, relates to the frequency with which such alternative bidding systems are to be used; while the Senate bill conditionally requires the Secretary to use methods other than the cash bonus bid with fixed royalty system for at least 50% of the total area offered for lease each year,[69] the House bill provides that alternate bidding systems will be used much less frequently. Under the House bill, the Secretary is conditionally required to apply systems other than bonus or variable royalty bidding only to 20% of the area offered for lease each year and is prohibited from applying them to more than 50% of the area.[70]

Historically, the Federal Government and the various States have followed the leasehold approach that has been common in the development of privately owned lands in the United States. The clear thrust of recent hearings in Congress and of S.9 is, however, that the United States Government should have a greater role in the management of its offshore resources, possibly even to the extent of developing them itself, as has been the trend elsewhere.[71]

The use of joint bidding in lease sales under the Outer Continental Shelf Lands Act has been gradually increasing in recent years; whereas, prior to 1972, it was unusual for more than one-half of the bids submitted in any lease sale to be jointly formed,[72] the presence of a majority of joint bids has become commonplace since that time. This trend is not necessarily undersirable, however, since joint bidding tends to increase competition when used by small companies which would otherwise not be able to participate in offshore lease sales. Offshore oil and gas leases are quite expensive[73] and there is almost always a substantial risk that the property being leased will not contain any petroleum. Joint bidding allows companies to spread their investment over a larger number of leases, thereby significantly reducing their risk. This seems to result in increased participation in OCS lease sales and higher lease sales.[74]

The use of joint bidding by the majors has, however, come under attach on the grounds that it has lessened competition and deterred the entry of the independents into OCS development.[75] While there is considerable question as to whether this conclusion is valid, joint bidding on Outer Continental Shelf leases among companies that produce more than 1.6 million barrels a day of crude oil, natural gas, and liquified petroleum was banned by the Department of Interior on September 29, 1975[76] and by Congress three months later.[77] The nine companies designated on the List of Restricted Joint Bidders, Texaco, Mobil, Socal, Exxon, Amoco, Arco, Shell, BP and Gulf, cannot bid with one another. They are, however, free to bid with other companies.

It should be noted that virtually every independent study, including two on which one of the authors participated,[78] has concluded that, notwithstanding extensive joint bidding, effective competition has existed in Outer Continental Shelf sales. It should also be observed that this type of restriction will predictably have no impact on the ability of the independent to make entry into U.S. offshore development. The costs of entry into such development are in all but a few parts of the United States so high as to exclude the small company, exclusive of bonus bidding considerations or any other preleasing costs. The only way the small companies have been able to effectively undertake work on the Outer Continental Shelf has been through joint bidding and where they have bid, they have bid effectively, sometimes jointly with the majors.[79]

Persons Who May Hold Leases

The Outher Continental Shelf Lands Act authorizes the grant of leases to "qualified" persons but contains no restrictions as to citizenship.[80] The regulations, however, restrict the holding of leases to citizens, resident aliens, or corporations of the United States or its states or territories.[81] Further, the Act authorizes any person with the approval of the Secretary of the Interior to conduct geological or geophysical operations in the Outer Continental Shelf[82] and the regulations do not contain any restriction in this regard. There is, therefore, relatively open competition for Outer Continental Shelf resources. Even though foreign nationals and corporations are not permitted to hold leases, they are free to use domestic corporations owned or controlled by them and have done so extensively.[83] British Petroleum through a subsidiary owns a major part of the North Slope Alaska reserves.[84]

The purpose of efficient resource management appears to be served by the existing system, which assures the federal government of legal jurisdiction over its Outer Continental Shelf lessees. This purpose is also served by the absence of any restriction on the number of acres that any operator can hold under lease. The class of entrants is, therefore, determined by economic interest in the resource offered. In this regard the Outer Continental Shelf Lands Act is clearly superior to the Mineral Leasing Act of 1920 with its individual acreage restrictions.[85]

Term

The Outer Continental Shelf Lands Act requires that oil and gas leases be for a period of five years "and as long thereafter as oil or gas may be produced from the area in paying quantities, or drilling or well reworking operations as approved by the Secretary (of the Interior) are conducted thereon" and carry a royalty of not less 12½%, which, as noted, in practice has been not less than 16-2/3%.[86] The Act gives the Secretary of the Interior the authority to prescribe rules and regulations for the "reduction of rentals or royalties"[87] and he has issued regulations authorizing the Director of U.S.G.S. to make such reduction "whenever he determines it necessary to promote development or finds that a lease cannot be successfully operated under the terms provided therein."[88] It appears that no application has yet been filled for such a reduction on outer continental shelf leases.

In addition the regulations of the Secretary permit the creation and transfer of "(c)arried working interests, overriding royalty interests, or payments out of production...without (any) requirement for filing or approval."[89]

It has been frequently suggested that the five-year primary term for oil and gas leases may be too short with respect to areas in which operations must be conducted on a short season basis, such as Alaska and other northern states.[90] It is also probable that exploratory drilling on the continental slope when and as the same is authorized by the Secretary of Interior may require greater time than that for areas in the shallower continental shelf.

Both S.9 and H.R. 1614 would meet this problem by authorizing a primary term of up to ten years where "necessary to encourage exploration and development in areas of unusually deep water or adverse weather conditions."[91]

Operations

The Outer Continental Shelf Lands Act authorizes the Secretary of the Interior to prescribe regulations determined "to be necessary and proper in order to provide for the prevention of waste and conservation of the natural resources of the Outer Continental Shelf, and the protection of correlative rights therein."[92] Pursuant thereto, the Secretary has promulgated regulations expressly made part of the Outer Continental Shelf leases authorizing U.S.G.S. supervisors to issue rules to prevent waste and "to govern the development and method of production of a pool, field, or area...to the end that all operations shall be conducted in a manner which will protect the natural resources of the Outer Continental Shelf and result in the maximum economic recovery of the mineral resources in a manner compatible with sound conservation practices."[93]

Before beginning operations, a lessee is required to file an acceptable plan for exploratory operations, which since the change of regulations following the Santa Barbara oil spill has included "features pertaining to pollution prevention and control."[94] Upon discovery the lessee must submit to U.S.G.S. for approval a development plan which also shows the location of proposed wells and detail therefor. Additionally each supervisor may require the lessee to drill other wells as may be reasonably required "in order that the lease may be properly and timely developed and produced in accordance with good operation practices."[95] Each supervisor is given authority to require tests to determine the identity and character of any formation, and is required to approve all well locations and the well spacing pattern for the proper development of the lease in question "giving consideration to such factors as the location of drilling platforms, the geological and reservoir characteristics of the field, the number of wells that can be economically drilled, the protection of correlative rights, and minimizing unreasonable interference with other uses of the Outer Continental Shelf area."[96]

In similar fashion the U.S.G.S. Supervisors appear to have been given the authority to fix production limits based upon whether the same would result in the maximum efficient rate of recovery ("MER"). The regulations authorize a Supervisor "to specify the time and method for determining the potential capacity of any well and to fix, after appropriate notice, the permissible production of any such well that may be produced when such action is necessary to prevent waste or to conform with such proration rules, schedules, or procedures as may be established by the Secretary."[97] "Waste" is defined to include "physical waste as that term is generally understood in the oil and gas industry (and) the locating, spacing, drilling, equipping operating, or producing of any oil or gas well or wells in a manner which causes or tends to cause reduction in the quantity of oil or gas ultimately recoverable from a pool under prudent and proper operations or which causes or tends to cause unnecessary or excessive surface loss or destruction of oil or gas."[98]

Historically, the Secretary of the Interior acquiesced in the application of state regulatory procedures with respect to drilling, spacing and prorationing or production restrictions which clearly resulted in inefficient management of the federal resources.[99] With the change in the petroleum supply situation in the United States over the past five years, this problem has been substantially eliminated. The dramatic changes which have taken place in the United States are illustrated by provisions in S.9 and H.R. 1614 which require production to proceed at a rate determined by the Secretary to be the maximum rate of production that is both safe and can be sustained without loss of ultimate recovery of oil or gas, unless the President decrees otherwise or the Secretary determines that another rate is "necessary."[100]

Other Uses

The Outer Continental Shelf Lands Act does not now authorize the construction of islands or other structures except for resource exploitation.[101] The 1958 Geneva Convention on the Continental Shelf has the same limited scope of authority.[102] The

Deepwater Port Act of 1974[103] does establish a mechanism for licensing and regulating the ownership and construction of "deepwater ports," defined as any facility beyond the territorial sea and used as a port or terminal for the loading, unloading or storage of oil.[104]

The Informal Composite Negotiating Text[105] that was issued by the Third U.N. Law of the Sea Conference on July 17, 1977 would authorize the coastal state in the proposed 200-mile exclusive economic zone "to construct and to authorize and regulate the construction, operation and use" of artificial islands, installations and structures, whether used for resource exploitation or not.[106] If adopted, it would be the first time such structures would be clearly authorized in international laws.

The Outer Continental Shelf Lands Act also fails to provide for the licensing or development of fresh water resources, living resources, salvage and treasury, and dredging and filling.

Elements of Current Policy

Introduction
In one form or another the United States has acknowledged policy objectives that are reflected in the Mineral Leasing Act, the Outer Continental Shelf Lands Act, other existing laws pertaining to petroleum and recently proposed legislation. Key among them are:

1. The establishment of an adequate and secure supply of petroleum;

2. The maintenance of a reasonable and predictable price for petroleum;

3. The maintenance of national security;

4. The maintenance of viable foreign relations;

5. Efficiency of resource utilization;

6. Protection of environmental quality;

7. The encouragement of free and effective competition;

8. The encouragement of private participation in resource development; and

9. The maximization of revenue to the federal government.[107]

There are, of course, many existing or potential conflicts among these objectives. Notable among these for its intensity is that between the encouragement of petroleum development and the protection of the environment.

Protection of Environmental Quality
Probably the single most significant product of the interest in environmental protection is the National Environmental Policy Act of 1969 ("NEPA").[108] The change in attitude toward resource management which NEPA implied has been summarized by one commentator as follows:

> "The idea incorporated in the policy statement of
> NEPA that valuable economic opportunity might in
> some instances be foregone in order to achieve an

environmental goal was a significant shift of
policy premises. Such a revolution in values
applied to government decision making would require an extraordinary mechanism to display and
weigh environmental effects of proposed actions,
just as economic effects had long been considered. [109]

NEPA requires that all "major Federal actions significantly affecting the quality
of the human environment" be accompanied by a detailed statement by the responsible
official on the environmental impact of the proposed action. [110] Included in the
statement must be an analysis of any adverse environmental effects or irreversible
and irretrievable commitment of resources which the implementation of the proposal
would cause or necessitate and a discussion of the relationship between local short-
term uses of man's environment and the maintenance and enhancement of long-term
productivity. [111] Alternatives to the proposed action must also be considered in the
statement. [111] Normally, the question of whether an environmental impact statement
would be required would depend upon all circumstances. [112] In the case of offshore
lease sales, [113] however, there is consensus by all that detailed impact statement must
be made. [113]

Other indications of the priority given to protection the environment are the Federal
Water Pollution Control Act, [114] the Coastal Zone Management Act of 1972, [115] S.9,
H.R. 1614, and recently enacted coastal legislation. The objective of the Federal
Water Pollution "is to restore and maintain the chemical, physical, and biological
integrity of the Nation's waters" and to this end the Act seeks the elimination of
the discharge of all pollutants into navigable waters of the United States by
1985. [116]

The Coastal Zone Management Act of 1972 declared it to be national policy "to preserve, protect, develop, and where possible, to restore or enhance, the resources
of the Nation's coastal zone." [117] The Act basically provides a mechanism for the
funding of coastal planning and its implementation by the various coastal states
within essentially the area of the three-mile territorial sea. The Act requires the
Secretary of Commerce, who administers it, to coordinate with other Federal agencies
and requires that they "to the maximum extent practicable...insure that any development project in the coastal zone of a state (is) consistent with approved state
management programs." [119] S.9 would require that, in preparing and maintaining the
gas and oil leasing program, the Secretary of the Interior consider the "policies
and plans promulgated by coastal States pursuant to the Coastal Zone Management Act
(or which) have been specifically identified by the Governors of such States as
relevant." [120] H.R. 1614 contains a similar provision. [121]

Both S.9 and H.R. 1614 also provide for the creation of an Offshore Oil Pollution
Compensation Fund which, together with individual owners and operators of offshore
facilities, would be strictly liable for losses resulting from oil discharges. The
owner and operator responsible for the damage would ordinarily be liable for the
first 35 million dollars in successful claims [122] and the Fund would be liable for
the balance. [123] The Fund would be financed in part by contributions of 3 cents per
barrel of oil produced under any lease issued under the Act. [124] S.9 would also
create a Fisherman Contingency Fund "for the purpose of providing compensation for
damages to commercial fishing vessels and gear and resulting loss of profits due
to activities of oil and gas exploration, production and development on the Outer
Continental Shelf." [125] H.R. 1614 contains a similar provision. [126] In addition,
in 1976 an amendment to the Coastal Zone Management Act was passed which created
the Coastal Energy Impact Program to provide "financial assistance to meet the needs
of coastal states and local government resulting from...activities involving energy
development." [127] Specified were needs "to provide new or improved public facilities
and public services which are required as a direct result of new or expanded Outer
Continental Shelf energy activity" [128] and of "prevention, reduction or amelioration

of any unavoidable loss...of any valuable environmental or recreational resource if such loss results from coastal energy activity."[129]

It should be noted that a major consideration in implementing the U.S. policy objective of encouraging private participation in the development of resources is the fact that no leasehold interest of an oil and gas lease in the United States whether issued by the federal government, a state government or a private party can be taken by a governmental entity without the payment of just compensation, which has been[130] construed to be the fair market value of the interest taken. This is a constitutional freedom guaranteed by the Fifth and Fourteenth Amendments and it is applicable to both U.S. and foreign nationals. It is thus clear that the cancellation of a federal oil and gas lease upon environmental or othergrounds as provided for in S.9[131] would require the payment of the fair value of the cancelled rights as of the date of their taking.

Conflicts Between the Government and the Petroleum Industry
The hearings on S.9 and related energy bills reveal a deep-seated suspicion on the part of a number of legislators that the private participation which has taken place with respect to Outer Continental Shelf petroleum resources has not been consistent with the public interest. The development of those resources has been dominated by the large multinational companies whose influence in both domestic and world markets is enormous. There is a perception that the world-wide interests of the companies control their attitudes in the development of domestic resources, including those in the Outer Continental Shelf. This has resulted in a number of proposals which in one form or another attempt to control the large companies. There has been proposed[132] legislation which would require vertical divestiture (breaking the major companies into producing, transporting, refining and marketing entities), horizontal divestiture[133] (prohibiting the companies from holding or acquiring non-petroleum energy resources), the disclosure and governmental approval for major foreign supply[134] arrangements and creation of a National Oil Company to undertake Outer Continental[135] Shelf drilling and, in one form, to deal in foreign supplies.

The large international companies have taken a hostile position to these proposals and the interest of the federal government in examining and controlling their operations. The absence of trust between them and the federal government has created an atmosphere in which the creation of an intelligent energy policy has been very difficult, if not impossible. The federal government, both administratively and legislatively, continues to view the companies as adversaries and, as a recent debate on deregulation shows, is politically inclined to attack any system which would give the companies an advantage, even a fair one.

With due regard to the magnitude, technological capabilities and managerial skills of the U.S. industry, it is readily apparent that a National Oil Company is not needed for functional purposes. On the other hand there is a clear-cut need to eliminate the confrontation between the large petroleum companies and the federal government and in this regard a government company might have a positive impact. If properly managed it could bring about a greater degree of realism to the federal government in policy-making and create a different relationship between government and industry. The cost involved in creating such an entity would be large but the political and economic benefits of reducing the government-industry hostility could be very substantial as well.[136] The fact that the United States is the last major petroleum producing or consuming nation without a national oil corporation gives support to those who favor this approach.

Conclusion

A number of mineral producing jurisdictions, both importing and exporting, have policy objectives similar to those of the United States and certainly the net impact of the U.S. system is not as dissimilar to theirs as might appear to be the case on first examination. Even though the basic contractual agreement used is a lease, rather than an operating agreement, net profits agreement, development contract or production sharing agreement, the federal government has increasingly been asserting its presence in the management of its energy resources, particularly those of the Outer Continental Shelf.. Even if S.9 is not enacted into law, it seems clear that this trend will continue. If S.9 is enacted into law as appears likely, it is certain that we will see a much stronger trend toward other arrangements, including those in which the federal government would have an operating interest.

The thrust of events in all parts of the world has maximized the position of the natural resource owner. Viewed in this context, it seems clear that the United States today is reassessing and broadening its knowledge and control of its energy resources. It is doing so, however, in the context of its own system of development under which private participation has traditionally been and is still being encouraged.

References

1. Offshore Oil Enters a New Era in the 70's, Offshore Mag. 61-63 (Jan. 1970).

2. Id. See also U.S. Dept. of Interior, United States Petroleum Production Through 1980, 14-18 (1968).

3. U.S. Dept. of Interior, Petroleum & Sulphur on the U.S. Continental Shelf 6 (Dec. 1969); See also 1 Nossaman, Water, Scott, Krueger & Riordan, Study of the Outer Continental Shelf Lands of the United States § 5.1 (1968) (hereinafter cited as Nossaman OCS Study).

4. Alaska Oil to Shake Up the Industry, The Oil & Gas Journal 99 (Apr. 20, 1970).

5. U.N. Sec.-Gen., Resources of the Sea, Part One: Mineral Resources of the Sea Beyond the Continental Shelf, U.N. Doc. E/449/add. 1, 14-17 (1968)

6. Id. at 7. 2 Nossaman OSC Study, supra note 3, at 5-A-53, 104,105.

7. Proc. No. 2667, 3 C.F.R. 67 (1943-1948 Comp.).

8. 3 C.F.R. 437 (1943-1948 Comp.).

9. See 4 M. Whiteman, Digest of International Law 752 (1965).

10. The definition of "Continental Shelf, Shelf Edge and Borderland" approved by the International Committee on the Nomenclature of Ocean Bottom Features is as follows:

> The zone around the continent, extending from the low water line to the depth at which there is a marked increase of slope to greater depth. Where this increase occurs, the term "shelf edge" is appropriate. Conventionally, the edge is taken at 100 fathoms (or 200 meters), but instances are known where the increase of slop occurs at more than 200 or less than

65 fathoms. When the zone below the water line in highly irregular, and includes depths well in excess of those typical of continental shelves, the term "continental borderland" is appropriate. 1 Y.B. Int'l L. Comm'n 131 (1956).

11. The controlling factor in determining whether an offshore area is continental shelf in the geologic sense should be whether the area is a natural extension of the continental land mass and is interior of the continental slope. See K.O. Emery, The Sea Off Southern California 5, 325 (1960); P. Keunen, Marine Geology 105, 158, 162, 339 (1950); F. P. Shepard, Submarine Geology 288-89, 425 (2d ed. 1963); Hearings on S. 1901, before the U.S. Senate Comm. on Interior and Insular Affairs, 83d Cong., 1st Sess. 210, 213 (1953).

=12. (1969) I.C.J. 3. See generally Krueger, The Background of the Doctrine of the Continental Shelf and the Outer Continental Shelf Lands Act, 10 Nat. Res. J. 442, 451 (1970) (hereinafter cited as Krueger, Background)

13. Krueger Background at 452.

14. 322 U.S. 19 (1947).

15. Id. at 38-39.

16. United States v. Texas, 339 U.S. 707 (1950); United States v. Louisiana, 339 U.S. 699 (1950).

17. 43 U.S.C.A. §§1301-15 (1964).

18. 43 U.S.C.A. §1311(a) (1964).

19. 43 U.S.C.A. §1301(b) (1964).

20. 43 U.S.C.A. §1301(b) (1964). See United States b. Louisiana, 363 U.S. 1(1960), rehearing denied, 364 U.S. 856 (1960); United States b. Florida, 363 U.S. 121 (1960).

21. 43 U.S.C. §§1331-43 (1964). The codification in the United States Code omits 13, 16 and 17 of the original Act. Sections 16 and 17 relate to appropiations and separability, respectively; 13 revoked Executive Order No. 10,426 which set aside the submerged lands of the continental shelf as a Naval Petroleum Reserve.

22. 43 U.S.C.A. §§1301-15 (1964).

23. 43 U.S.C.A. §1332(a) (1964).

24. S. Rep. No. 411, 83d Cong., 1st Sess. 2, 4-7, 211-24 (1953); H.R. Rep. No. 413, 83d Cong., 1st Sess. 2, 6-7 (1953). See Stone, United States Legislation Relating to the Continental Shelf, 17 Int'l & Comp. L. A. 103, Shelf Lands Act; Key to a New Frontier, 6 Stan.L.Rev. 23, 26 (1953).

25. 43 U.S.C.A. §1331(a) (1964). Cf. The language of the Truman Proclamation: "(T)he United States regards the natural resources of the subsoil and sea bed of the continental shelf . . . as appertaining to the United States, subject to its jurisdiction and control." Pres. Proc. No. 2667, 3 C.F.R. 67 (1943-1948 Comp.), (1945).

26. See Memorandum Opinion (M36615/94127/61) from Assoc. Solictor, Dep't Interior to Director BLM (May 5, 1961); Barry, The Administration of the Outer Continental Shelf Lands Act. 1 Nat. Res. Law (No. 3) 38, 46 (1968).

27. S. Rep. No. 133, 83d Cong., 1st Sess. 2 (1953), stated that the Act was to give "the weight of statutory law to the jurisdiction asserted by the proclamation of the President of the United States in 1945."

28. 43 U.S.C.A. §§1337, 1341 (1964).

29. See, e.g., 43 C.F.R. §§3382.1 et seq. (1964).

30. C.F.R. §§250.14, 250.34 (1954). See 1 Nossaman OCS Study, supra note 3, SS4.9, 4.36.

31. See 1 Nossaman OCS Study, supra note 3, at 397-402

32. Id., §§4.12-4.16, 11.5-11.8.

33. The spill was caused by the blowout of one of the wells being operated by Union Oil Company under a lease with the federal government and resulted in the release of more than 250,000 gallons of oil into the Santa Barbara Channel. See generally, Walmsley, Oil Pollution Problems Arising out of Exploitation of the Continental Shelf: The Santa Barbara Disaster, 9 San Diego L. Rev. 514 (1972).

34. 43 C.F.R. §3301.4 (1976).

35. Id.

36. See Cal. Pub. Res. Code S6873.2 (West Supp. 1977).

37. 43 C.F.R. §3301.4 (1976). New rules will "set up an expanded OCS oil and gas information program" and give the governors of coastal states "an opportunity to review and comment upon offshore decisions affecting then, "The Oil and Gas Journal, p. 63, Oct. 3, 1977.

38. S.9 (Jackson), 95th Cong. 1st Session (1977). See 123 Cong. Rec.S 11984, et seq. (daily ed. July 15, 1977). A similar bill was passed by the Senate in in 1975 (see 121 Cong. Rec. S 14,362 (daily ed. July 30, 1975) and by the House of Representatives in 1976 (see 122 Cong. Rec. H 7514 (daily ed. July 21, 1976) and 122 Cong. Rec. H 8021 (daily ed. July 30, 1976)), but it was never reported out of the Conference Committee which had been appointed to resolve the differences in the two versions.

39. S.9, supra note 38, §206.

40. Id. at §511.

41. Id. at §26.

42. 16 U.S.C.A. SS1451-64 (1974, Supp. 1977).

43. S.9, supra note 39, §18(F).

44. Id. at §19.

45. Id. at S18(a)(2)(F).

46. 33 U.S.C.A. §§1501-24 (Supp. 1977).

47. 42 U.S.C.A. §§4321-47 (1977).

48. H.R. 1614, 95th Cong., 2d Sess. (1978). See 124 Cong. Rec. H 602, et seq. (daily ed. Feb. 2, 1978).

49. See note 224, supra, and accompanying text.

50. H.R. 1614, supra note 223, §207(g).

51. S.9, supra note 223, §205(a)(3).

52. 43 U.S.C.A. §1337(b) (1974).

53. U.S. Dept. of Interior, 94th Cong., 2d Sess, Joint Bidding for Federal Onshore Oil and Gas Lands and Coal and Oil Shale Lands 19 (Comm. Print 1976) (hereinafter cited as Interior Report).

54. U.S. Dept. of Interior, Call for Nominations and Comments on Area for Oil and Gas Leasing (Tentative Sale #48), July 16, 1976.

55. This is in large part because of the trend in recent years towards larger tracts and higher prices per acre. Id.

56. See generally, Kreuger, An Evaluation of the Provisions and Policies of the Outer Continental Shelf Lands Act, 10 Nat. Res. J. 763-85 (1970) (hereinafter cited as Krueger, Evaluation).

57. 1 Nossaman OCS Study, supra, note 188, at §11.24

58. S.9, supra note 39, §205(b)(1).

59. H.R. 1614, supra note 48, §205(b)(1).

60. 43 U.S.C.A. §1337(a) (1964). See 1 Nossaman OCS Study, supra note 3, §4.30.

61. 43 U.S.C.A. §§1337(d)-(e) (1964). See 1 Nossaman OCS Study, supra note 3, S4.31.

62. 43 U.S.C.A. §§1337(a)-(b) (1964). In any case, a flat royalty of not less than 12-1/2% would be required to be paid in addition to such other terms as are fixed by the Secretary.

63. 1 Nossaman OCS Study, supra note 3, at 614.

64. Cal. Pub. Res. Code §6827 (West Supp. 1977). See generally 2 Nossaman OCS, supra note 3, at 12-A-8, 9, 16, 17, 18.

65. See generally 2 Nossaman OCS Study, supra note 3, at 12-c-75, 76.

66. See generally, Id. at 12-c-58, 59.

67. S.9, supra note 38, §205(a)(1); H.R. 1614, supra note 233, §205(a)(1).

68. H.R. 1614, supra note 48, §205(a)(1)(J).

69. S.9, 30 U.S.C.A. §201(b)(3) (Supp. 1977) 205(a)*8)(a).

70. H.R. 1614, 30 U.S.C.A. §205(a)(5)(C). Advance royalties cannot be deducted from production royalties if they have been used to reduce such royalties in a prior year; nor can an advance royalty paid during the initial twenty year term of a lease be used to reduce a production royalty after the twentieth year of a lease.

71. See House of Representatives, Select Ad Hoc Committee an Outer Continental Shelf Report on H.R. 1614 - Outer Continental Shelf Lands Ace Amendments, August 29, 1977; Cong. Rec. § 11929, et seq., re §.9 (daily ed. July 15, 1977).

72. Most winning bids are also joint bids, as is indicated in the following table printed in Interior Report:

Joint Bidding in OCS Lease Sales

Sale Number	Date	State	Percent Joint Bids	Percent High Joint Bids
1	10/31/54	La	18	19
2	11/9/54	Tex	24	32
3	7/12/55	La	39	29
4	7/12/55	Tex	24	18
5	5/29/59	Fla	96	96
6	8/11/59	Ka	29	37
7	2/24/60	La	42	54
8	2/24/60	Tex	15	17
9	3/13/62	La	24	20
10	3/16/62	La	27	22
11	3/16/62	Tex	0	0
12	10/9/62	La	42	44
13	5/14/63	Cal	14	7
14	4/28/64	La	14	22
15	10/1/64	Ore	70	68
16	10/1/64	Wash	58	64
17	3/29/66	La	28	35
18	10/18/66	La	37	46
19	12/15/66	Cal	74	100
20	6/13/67	La	38	48
21	2/6/68	Cal	74	75
22	5/21/64	Tex	49	32
23	11/19/68	La	45	69
24	1/14/69	La	40	55
25	12/16/69	La	28	25
26	7/21/70	La	19	26
27	12/15/70	La	47	51
28	11/4/71	La	36	46
29	9/12/72	La	52	68
30	12/19/72	La	77	73
31	6/19/73	Tex/La	80	92
32	12/20/73	Miss/Ala/La	62	63
33	3/28/74	La	68	73
34	5/29/74	Tex	60	61
35	12/11/75	Cal	45	43
36	10/16/74	La	62	57
37	2/4/75	Tex	52	43
38	5/28/75	La/Tex	53	52
38a	7/29/75	La/Tex	54	55

73. The average price paid for an OCS lease is $5.88 million. Interior Report.

74. Id. at 19-35.

75. Gaskins, Darius and Barry Vann, Joint Buying and the Seller's Return--the Case of OCS Lease Sales (1975), discussed at Interior Report, supra note 53, at-15-35.

76. 40 Fed. Reg. 45172 (1975), codified in 43 C.F.R. §3302.3-2 (1975). The department recommended in 1976 that the ban be extended to cover onshore oil and gas leasing, as well (Interior Report, supra note 53, at 53), but no action has yet been taken on the recommendation.

77. Energy Policy and Conservation Act, 89 Stat. 871, §105 (1975), codified in 42 U.S.C.A. §6213 (1977).

78. See Nossaman OCS Study, supra note 3, §§8.15-8.22, 11.24, see R. Krueger, The United States and International Oil: A Report for the Federal Energy Administration on U.S. Firms and Government Policy 98 (1975) (hereinafter cited as FEA Study), Ch.1.

79. See 1 Nossaman OCS Study, supra note 3, §8.13-8.14.

80. 43 U.S.C.A. §1337(a) (1964).

81. 43 C.F.R. §3300.1 (1976).

82. 43 U.S.C.A. §1340 (1964).

83. See 1 Nossaman OCS Study, supra note 3, Tables 8-3.

84. See FEA Study, supra note 78 at A-26.

85. 41 Stat. 437 (1920), as amended, 30 U.S.C.A. §184 (1971).

86. 43 U.S.C.A. § 1337 (1964).

87. 43 U.S.C.A. §1334(a)(1) (1964).

88. 30 C.F.R. §250.12(e) (1977).

89. 43 C.F.R. §3305.2(a) (1976). The resource misallocation problem which inherently results from a fixed royalty is compounded by the lessees' freedom to create such additional burdens on production. 1 Nossaman OCS Study, supra note 3, §11.43.

90. 1 Nossaman OCS Study, supra note 3, §11.45. See also Commission on Marine Science, Engineering and Resources, Our Nation and the Sea: A Plan for National Action 126-27 (1969).

91. S.9, supra note 69, §205(b)(2); H.R. 1614, supra note 48, §205(b)(2).

92. 43 U.S.C.A. 1334 (a)(1) (1964).

93. 30 C.F.R. §250.11 (1977).

94. 30 C.F.R. §250.34 (1) (1977).

218

95. 30 C,F,R, §250.33(b) (1977).

96. 30 C.F.R. §250.17 (1977).

97. 30 C.F.R. §250.16 (1977).

98. 30 C.F.R. §250.2(h) (1977).

99. See Krueger, Evaluation, supra note 56. at 797.

100. S.9, supra note 69, §204(g); H.R. 1614, supra note 70, §204(f).

101. See Krueger, Background, supra note 12, at 476-78.

102. See Id., at 476-78.

103. 88 Stat. 2126 (codified at 33 U.S.C.A. SS1501-24 (Supp. 1977) and 43 U.S.C.A. §1333(a) (Supp. 1977).

104. 33 U.S.C.A. §1502(10) (Supp. 1977).

105. Third United Nations Conference on the Law of the Sea, RCNT Informal Composite Negotiating Text, U.N. Doc. A/CONF. 63/WP. 10 (July 15, 1977).

106. Id., Art. 56, Para. 1.

107. See FEA Study, supra note 78, Ch. 2.

108. 42 U.S.C.A. §§4321-47 (1977).

109. Dreyful & Ingram, The National Environmental Policy Act: A View of Intent and Practice, 16 Nat. Res. J., 243, 245 (1976).

110. 42 U.S.C.A. §4332(c) (1977).

111. Id.

112. See Hanly v. Kleindienst, 471 F.2d 823, 836 (2d Cir. 1972).

113. See U.S. Council on Environmental Quality, OCS Oil and Gas - An Environmental Assessment (1974); Commerce Comm., 93rd Cong., 2nd Session, Outer Continental Shelf Oil and Gas Development and the Coastal Zone 37 et seq. (1974).

114. 33 U.S.C.A. SS1251-1376 (Supp. 1977).

115. 16 U.S.C.A. §§1451-64 (1974), Supp. 1977).

116. 33 U.S.C.A. §1251 (Supp. 1977).

117. 16 U.S.C.A. §1452 (1974).

118. 16 U.S.C.A. §1456(a) (Supp. 1977).

119. 16 U.S.C.A. §1456(c)(2) (1974).

120. S.9, supra note 69, §18(a)(2)(F).

121. H.R. 1614, supra note 127, §18(a)(2)(F).

122. S.9, supra note 69, §308; H.R. 1614, supra note 70, §304.

123. S.9, supra note 69, §309; H.R. 1614, supra note 70, §307.

124. S.9, supra note 69, §310; H.R. 1614, supra note 70, §302.

125. S.9, supra note 69, §§402, 404 et seq.

126. H.R. 1614, supra note 70, §30.

127. S. 586, 94th Cong., 2nd Sess. (1976), Cong. Rec. H 6683, H 6688 §308(a)(1).

128. Id. at §308(a)(B)(i).

129. Id. at §308(a)(C).

130. ". . . (N)or shall private property be taken for public use without payment of just compensation." U.S. Const. Amend V. This provision was applied to the taking of a leasehold interest in Almota Farmers Elev. & Wrhse. Co. v. United States, 409 U.S. 470 (1973). See also Lemmons v. United States, 496 F.2d 864, 873, (Ct. Cl. 1974).

131. S.9, supra note 69, §204.

132. S.2387, 94th Cong., 1st Sess.

133. H.R. 3370, 95th Cong. 1st Sess.

134. Such a proposal was considered during August, 1974 by the Subcommittee on Multinational Corporations of the Senate Committee on Foreign Relations. (93rd Cong., 2d Sess.). See FEA Study, supra note 78, Ch. 6.

135. Proposed Consumer Energy Act of 1974, SS301-02. Hearings, U.S. Senate Commerce Comm., 93rd Cong., 2d Sess. Pt. 4, 1409 (April 22-23, 1974). See FEA Study, supra note 78, at 251.

136. See The Governance of International Oil: The Proper Roles for Industry and the U.S. Government, Seven Springs (Yale) Report, Sect. VI, Sept. 7, 1977.

16

THE REGULATORY NIGHTMARE: CATCH 22 !

by

Donald B. Bright

Port of Long Beach, California

The institutions of American business are creaking from an overburden of regulatory restrictions that have become so enormous that it is virtually impossible to respond. It is a situation where the process has become the god, and the product--an environmentally safe project-- almost forgotten. The reasons for overregulation are illustrated by the problem of energy need (use) versus conservation.

Energy sources and energy uses are primary concerns today. So is saving the environment. These concerns are not mutually exclusive. Let's spend a moment developing a common frame of reference considering the philosophical and cultural aspects of energy use in the U.S.

One of the most profound changes occurring since the beginning of life has been the way man has modified and to some extent, "controlled" his environment. In essence, our existence on the earth has affected it, changed it as well as reduced its resources. Simultaneously man is both the peril and promise. Recent events have caused us to pause and ask: Can we live on the earth responsibly?

Man is unique -- different from other organisms because of his great power to improve as well as exploit. In today's life there is a tendency for man not to use his muscles -- we build machines to do our labor. These are our energy-saving devices-- but, in reality, they are energy-using devices. For example, every person in the U.S. uses 8400 KWH/YR of electricity or about the equivalent of 10 servants. Electric machinery allows one person to do the work of about 683 persons. This usage is still increasing, for we use 16 times as much electricity today as in 1920. Also, total gross energy consumption doubled between 1950 and 1970--and will likely double again by 1990 unless we alter our patterns of energy use.

For some, the problem seems simple--"energy is essential to the health, safety and well-being of all persons." Accordingly, we must take every opportunity to insure an appropriate economic strategy between supply and demand. On the other hand, this issue is equally simple to others, that is, "we waste about 50% of our energy"; therefore, we need only to improve our technology, alter our lifestyle and relax. But our uses of energy are based on socio-economic demands--demands which dominate our lives. Thus. to suggest that we switch our lifestyle, immediately, today, is not reasonable. For example transportation of people and freight consumes 20% of the total energy used (primarily in the form of petroleum); 18% is used for space heating of homes and commercial establishments; 42% for industrial uses (e.g., to provide processed steam, direct heating, electric drive, etc.); the remaining 15% is used commercially and residentially for water heating, air conditioning, refrigeration, cooking, lighting, small appliances, etc. In essence, our evolving technologies are characterized by an increasing rate of energy intensiveness. This has been essential to the economy, i.e., one of the best measurements of the state

of the economy is the level of per capita energy consumption. I am not an econo-
mist, but it does seem clear that we will need several years of readjustment to
energy needs or costs-- and these changes also will demand adjustments in product
technologies.

Now, what about the environment? Environmentalism isn't new. Environmental control has been a prime goal of man since his very beginning. Controls such as
cultivation and altered water flow were aimed at "overcoming" the environment.
Today, however, the scene is compound to a critical point by an everincreasing
population, dwindling of pristine areas and available resources, and multiple-use
demands.

Today, the public wants to redevelop, reclaim, restore, protect, preserve and
always maintain balance. However, all six of these processes demand new tools,
not only for accomplishment, but for decisions leading to accomplishment. Accordingly, the cultural phenomenon called law is called on with increasing frequency to
regulate conflicting constituencies. But, this further compounds the conflict
since the law tends to be the baliff for the ruling power of the "moment."

The forces of "development" have not shown an aptitude for creating an environment
that can be stable, self-sustaining and/or renewing. Rather, as congestion expands
and the rate of expansion increases, the environment in both a specific and general
sense has undergone tragic defacement. The most obvious results have been confusion, conflict, hastily imposed rules and regulations, and as a result, inadequate
positive action.

Armed with evidence of this confusion,environmentalists have pressed hard for
blanket prohibitions, moritoria, complex and confusing almost - prohibitory rules
and regulations,where all changes are precluded except to carry out "desirable"
restoration. Yet, such prohibitions alone do not represent viable solutions. In
point of fact, they are just as destructive because of direct disturbance of economic stability, general cultural customs and governmental management protection
procedures.

There also is a common myth, that simple terms and conditions can be imposed on
any project so as to protect the environment. There is a "feeling" that with enough
studies and reports and accumulation of evidence, that all will be well. Yet,
these are erroneous assumptions that deny and ignore the complex workings of our
environment. This inordinately confusing situation is likeable to Peter's placebo:
i.e., "An ounce of image is worth a pound of performance!" Additionally, the situation is like Matsch's law: "It is better to have a horrible ending than to have
horrors without end." Just to round off these comparisons we should include Fair-
fax's law: i.e., "Facts which, when included in the argument, give the desired result, are fair facts for the argument."

Having reviewed the philosophy of energy use and the related concern for the environment, let's turn to the existing solution -- the evolution of regulatory
agencies as an out-growth of expanding governmental processes and public concern.

In spite of the increasing frequency with which the law is used to regulate conflicting constituencies, it is axiomatic that we still are left only with one prime
tool: regulation. Unfortunately, the regulatory process has been fully accepted
by the public as necessary. Simultaneous with this acceptance has been a failure
to recognize the inconsistencies, the fraility and the demagoguery of regulations.
But do not misunderstand, I am not implying that the regulatory process must be
obliterated, nor am I implying that regulations are inherently bad; rather, the
implication is that far too many regulations are being created for the benefit of
the "regulator" rather than the "regulatee."

Why has the regulatory process gone so far askew, why has it evolved into a processor's delight and an applicant's nightmare? The basic problem centers around the development of regulations as an "after thought." The majority of the regulations controlling our lives have evolved as a consequence of the determination of a "major" problem. Because the problem was determined after the original development, the regulation that evolved is in the form of a "band aid", it seldom is corrective surgery. It is this after-the-fact approach which has produced the greatest problems in the regulatory process.

Because of this after the fact approach, because of the ever increasing number of regulatory agencies and because of the sheer human tendency to obtain and administer control, the productivity of the regulatory process is far less than it should be.

A review of several local, state and federal regulatory agencies indicates that certain problems are quite common. The following illustrate this statement:

1. Absence of Coordinated Responses to Common Issues

Federal, state and local agencies often fail to work toward common goals on behalf of the public. Examples here are numerous:

a) EPA and ACE had great difficulty agreeing on section 404 rules regarding dredge spoil disposal. At the outset, what evolved can be categorized as a "best guess" regulation. There were no definitive data for establishing normal background levels of contaminates at dredge disposal sites and no firm methodology for addressing levels of pollutants in dredge spoil. The permitting process thus became a research process resulting in considerable delay and greatly increased costs to the project proponent. Concomitantly the credibility of the regulatory agencies was reduced. It is even questionable whether or not there was a net benefit to the environment.

b) The various air emission regulatory agencies have had considerable conflicts on the collection and analysis of data. This is clearly illustrated by the disagreement between EPA and the California Air Agencies over whether or not Los Angeles was/is in violation of federal SO_2 standards. Often this is not a regulatory issue, rather it is an issue of great and grave political gamesmanship.

c) The various regional units of the same state agency often enforce the laws, rules, procedures and/or guidelines quite differently. Regionally specific decisions are essential. For example, what is appropriate for "Redwood City" would not always be appropriate for "Malibu" and vice versa. Yet common principles should be enforced fairly, evenly throughout the state.

d) The problems related to outer continental shelf (OCS) decisions were not clearly/effectively addressed before-the-fact by California State regulatory agencies with respect to lease sale #35 off the coast of southern California. In point of fact, at that time the State had not clearly defined energy policy nor any agency/co-agency task force to coordinate/orchestrate state efforts. Now, as a consequence, oil companies trying to comply with the requirements of the lease sale are dealing with state agencies to resolve issues and principles while at the same time the State is challenging the validity of the OCS lease in the court of appeals. A decision in favor of the State would serve to void the leases. Therefore, developers are proceeding at considerable risk.

Enough problems. It is more important to be positive--like the peace corps in their advertisement with a glass of water which is half full--not half empty!

Clearly, amid all the "furor" about eco-ethics, environmental controls, the maze of regulatory processes and related documents, two facts have emerged:

- That the concern for reasonably preserving the environmental quality has diminished and has therefore simultaneously increased in value to the average citizen.

- That future protection of the environment is going to be costly and that government, as the public's trustee, must reasonably protect the environment for the public while simultaneously supporting the publics revered institutions and processes.

Considering these two facts, what are the needs for the future?

Firstly, because regulatory constraints will continue to increase, the public must hold firm for better, more efficient procedures coupled with viable public participation; also, these procedures must be based on a balanced review of all impacts/problems associated with a given project.

Secondly, adequately trained professionals quickly must be brought on stream to bridge the gaps between academic pronouncements, bureaucratic dicta, and public needs.

Thirdly, some few national goals with the full support of those espousing the development--as well as the conservation--ethic, must be determined and established to serve as the foundation for the evolution of appropriate procedural/government mechanisms.

Fourthly, corporate, utility and public managers must cease to make development decisions as if they operated in a "black box". All too often they become totally frustrated after-the-fact when the carefully negotiated agreements/leases are denied because of non-compliance with existing rules and procedures. A forthright out-front consideration of applicable rules and regulations will go a long way toward evolving projects which are environmentally feasible as well as economical.

Lastly, we must overtly pursue a course of endeavor regarding the environment which instills a high degree of individual responsibility. This last point represents the hinge of fate for the survivorship of man. The desire to be regulated is strong, the need for regulation is clear but when regulations become a goal in themselves, those that are being regulated rebel.

Four key conclusions evolve from this review:

- The regulatory process is well established and will not "go away".

- The regulatory review and management processes must be melded to achieve "best reasonable" resolution of divergent issues.

- The regulatory review must be based on a composite analyses of all key issues.

- The regulatory process must be effectuated at the start of projects--long before the design engineer puts "lines" on a schematic drawing.

2. Narrow Responsibility

Often through bureaucratic refinement, the scope, aim, and purpose of a given agency is focused on a narrow responsibility. This position is most commonly defined by the statements that "budgetary limits preclude our doing more" or "that does not fall within our jurisdiction". Simply stated, this means that any applicant undergoing review for approval of a project can have his project denied on the basis of a single limiting factor rather than a balanced composite analysis of all the impacts and potential mitigations.

3. Limited or Narrow Scope of Power

Many regulatory agencies evolved early in the process and have not been "upgraded" to meet recent demands. For example, the California Department of Fish and Game is often erroneously faulted because they have not assumed full, effective trusteeship of California's marine resources, and because they have not accomplished essential research related to maintenance of these resources. Yet, by the state constitution, they are not eligible for general fund monies, so, they must rely almost totally on revenue obtained from selling permits, licenses, etc. No agency, however herculean the effort, can do the job without appropriate resources.

4. Cumbersome Procedures

Often, the overcoming of agency enertia is greater than accomplishing the project. The extensive application forms, the overlapping review (7-15 years) for decisions on nuclear facilities is a prime example. The slow development of California Environmental Quality Act, (CEQA) guidelines, essentially three years after passage of the act, also illustrates this problem. Further, recent examples from air quality regulations include: new source review, potential significant deterioration, state implementation plan, and so forth.

5. Being Last is "Golden"

The philosophy has evolved, and become golden, that a "good" regulatory agency sits in judgement on the decision of other agencies, i.e., the last to approve is the best or the most effective.

6. Overlapping Jurisdictions and/or "Passing On" the Responsibility to Another Agency

This is a cause and effect type action, based on an absence of coordinated response to common issues, inflexibility, limited scope of power, cumbersome procedures and an after-the-fact approach. Emerson said it better, i.e., "...consistency is the hobgoblin of little minds..."

Now, what have these indigenous regulatory problems produced? In general the results have been lengthy procedures which lead to approval with complex conditions which:

- Only large corporations with enormous economic incentives can accomplish.

- Only can be accomplished by the first few applicants using the process, e.g. air emission trade-off sources which are quickly exhausted.

- Are tantamount to denial.

17

OCS IMPACT ON RECREATION - THE SOUTHERN CALIFORNIA VIEW

by

Susan H. Anderson
Sea Grant Marine Advisory Services

The impact of outer continental shelf development on recreation varies considerably depending, not so much on the region, but rather the characteristics of the community near which the outer continental shelf development takes place.

In most instances there are some positive benefits to recreation. The rigs, for instance, provide artificial reefs to enhance fishing, although at the same time they may also provide obstacles to navigation. The OCS lease revenues are contributed to the Land and Water Conservation Fund. In the years 1969 to 1974 OCS revenues made up 71% of the total in the Land and Water Conservation Fund to the tune of $1,136,397,707.00. To give you an example of how these revenues have returned to recreation, in 1974 Santa Barbara and Ventura Counties alone received $5,390,033 for recreational projects.

One of the most common impacts discussed is the population boom created by tremendous new development of any kind and, in particular, OCS development, because of the specialized technology that requires bringing in many people from out of the area. Population booms have had a high impact in areas like Alaska, some of the North Sea towns, and in the Gulf Coast of the United States, but to date population booms have had a limited impact on southern California. A population boom might occur in the Ventura and Oxnard area if future leases are made between the Channel Islands and San Nicholas. However, because of the nature of development that already exists on the southern California coast in this area, the heaviest impact of OCS development occurs with a blowout or a spill.

My remarks in this paper will be directed towards that impact as perceived in the Santa Barbara blowout of 1969. In that blowout, 3,250,000 gallons of oil were spilled. Following the disaster, local communities as well as private organizations attempted to assess the damage.

The beach clean-up projects and related oil collection and oil-well control cost companies $10,487,000.00. The portion of this clean up, monitoring, and inspection that accrued to government cost $639,000.00. Biologists, in reviewing the dead organisms caused by the spill, estimated that replacement costs for those organisms was over $10 million. Since, in fact, the organisms would not be replaced by purchase from scientific laboratories, this $10 million represents a value rather than a cost. The cost to replenish the sand that had been spoiled by the oil was $346,370.00. Beaches that were in public domain, both state and local, diminished in property value and yielded interim rent and utility losses between 1969 and 1974 of between $6.9 million and $8.9 million because of sand loss and degradation of the environment caused by the blowout.

Sales tax losses were incurred or at least the sales taxes were diverted to other areas of the state. $17 million were lost or diverted in state taxes. The County of Santa Barbara incurred a $1.6 million tax loss including loss in property taxes. The City of Santa Barbara realized a 3.6 million loss including $600,000.00 in bed taxes. The City of Carpenteria experienced a loss in taxes of $566,000.00. In addition, there was a diminution of public services to local communities of undetermined value, because of redeployment of personnel for clean-up, as well as because of lower tax revenues.

The value of the lost beach experience was determined to be over $3 million with a minimum of 744,000 fewer visits to the beach in the one to two years immediately following the spill. Boats could not get out of the Santa Barbara Harbor for several months because of oil containment booms. The harbor holds 600 boats. In using the estimate of the State Department of Navigation and Ocean Development that the boats in southern California's coast are used an average of 27 days a year, we can determine that this was a loss of roughly 16,000 recreational boating days.

Sportfishing out of Santa Barbara declined from 10,000 anglers in 1968 to 2,000 in 1969 on partyboats alone. The counts for the number of private boats going out to catch fish are not available.

Except for beach use, these figures do not include the value of lost recreation days, nor do they include loss of wildlife experiences for recreationists, such as photography, painting, and bird watching.

Most of these losses discussed were not permanent losses. Santa Barbara today realizes a high level of economic return from recreation in that community. But the fact that the loss was only temporary does not mean that it was not significant, both economically and psychologically, nor that no improvements are needed to help alleviate such problems in the future.

Since 1969 there has been greater attention paid to technology to decrease the probability of blowout but more attention is needed to oil containment practices and recovery techniques. There is a need for increased response capability, a need to have equipment more readily available. The present liability mechanisms are inadequate.

Following the 1969 spill, the oil companies in the area formed a consortium called "Clean Seas Incorporated". This consortium responds to member-related spills or seepage only, unless they are called in by the Coast Guard. The focus of their efforts is to limit their own liabilities. Even with Clean Seas Inc. there is only a limited boom length available. If spills are not contained, the loss of aesthetics and resources is not irreparable but it is significant. Existing Federal Government emergency funds for mobilizing effort to clean up a spill when the source is unknown does not usually react fast enough. There is a need perhaps for stronger governmental intervention, monitoring and supervision.

The Los Angeles County Department of Beaches has put into effect a monitoring program for spill protection. But this occurs too late to prevent damage. It can only help them to some extent in limiting the damage.

The needs of recreational users of the marine environment for protection from accidents related to OCS development are similar to those of all of us. We need provision of adequate clean-up facilities on the nearest available land base and, where necessary, on support barges adjacent to drilling facilities. The first responsibility rests with the oil companies carrying out the development.

In what ways might we ask the oil companies to mitigate the effects of their de-

velopment in a manner that would be of benefit to the recreational community?
If a base of operation is developed, for instance on the Channel Islands, it
might be appropriate to have the oil companies include facilities for a harbor of
refuge marina for boaters. This would not be the first time that a marina was
built as a mitigation effort. There is at least one precedent in one of the North
Sea facilities where, following the development of an offshore platform on a low
coastal area which had to be flooded in order to float the platform out ot sea,
a full recreational marina and park was built in the devastated area by the developing oil company.

Recreational islands might be built around the rigs similar to the Thums Islands
in Long Beach with dedication for recreation. A sand replenishment program might
be included in OCS development plans. We might also seek coastal park acquisition
and development as a requirement in mitigation for potential damages.

Funds from the Coastal Environmental Impact Program might be requested for use by
local government containment programs and for local clean-up facilities. In particular, recreational users might push to have those funds used for longer booms
to be more effective in beach clean-up for long stretches of sandy beach.

The greatest impact that outer continental shelf development has on recreation in
a heavily populated area is the temporary and long-term loss of aesthetics and
recreational facilities and thereby recreational opportunities. Obviously, prevention of oil spills or blowouts is the first line of defense, but in the event
of an accident the oil companies, the local governments, and the managers of recreation facilities need to be prepared for clean-up procedures. Based on history,
it seems inevitable that recreational facilities, as well as other coastal properties, will be damaged. It is up to recreation managers and users of the coast to
make their demands for mitigation efforts by the oil companies in order to ensure
the best available options and increased alternative areas for recreation in the
event of a spill.

18

EFFECTS OF A POWER PLANT EFFLUENT ON
INTERTIDAL ORGANISMS AT HUMBOLDT BAY, CALIFORNIA

by

Joseph S. Devinny
University of Southern California

Summary

The heated water discharged by coastal pow r plants alters local ecosystems. Effects
on an intertidal biological community at Humboldt Bay have been measured. Species
composition in the lower intertidal macroalgal community changes by about 6% for
each centigrade degree rise in average temperature. Similar but smaller changes
occur in the lower intertidal animal community. In the high intertidal, the water
temperature changes have little effect on either algae or animals.

Introduction

All large-scale electrical generating stations, whether coal-fired, oil-fired, or
nuclear, share the problem of waste heat disposal.

Coastal power plant siting provides an economical solution: once-through cooling of
the condensors with seawater. The resulting thermal discharges, however, change
local ecosystems. Heat shock, mechanical abrasion, or added chemicals kill or in-
jure some organisms entrained in the flow. Discharges alter pelagic communities in
the area because some species are repelled and others are attracted by the warm
water. Where the effluent contacts the bottom or shore, the temperature change
modifies the community of sessile organisms. Where the warmed water contacts a
rocky shoreline, it affects a particularly valuable ecosystem. This article presents study of an example of thermal effects on rocky intertidal flora and fauna.

The Humboldt Bay Power Plant discharges water through a surface canal at the study
site (Fig. 1). Substrate at the canal mouth is sandstone rip-rap relatively uniform in slope and boulder size. Extensive descriptions of the site and associated
biotic and abiotic data were published by Adams (1975) and the Pacific Gas and
Electric Company (1973). This existing data is analyzed here to provide a general,
community-based measure of outfall effects.

Methods

Adams established transects at fourteen places near the outfall. Two were at the
outfall; others were symmetrically placed 10, 30, 50, 150, 500 and 1000 m to the
northeast and southwest. Records included the biological community within about
1 m on each side of each transect. Data were recorded separately for four different tidal levels: 0-.61 m (0-2 ft), .61-1.5 m (2-5 ft), 1.5-2.1 m (5-7 ft), and
2.1-3.0 m (7-10 ft) (measured above mean lower low water, U.S. Nautical Datum).

Species lists were prepared for this analysis from data collected by Adams in
February and May of 1972 as follows:

1. Separate lists were complied of all animal species and all plant
   species found at each transect and tidal level in both surveys.

2. Lists for two sites were eliminated because the rock substrate was
   covered by sand during one of the surveys, reducing the comparability
   of the data.

3. Lists for the highest tidal level were eliminated because few species were present.

The lists were thus presence and absence data for algae and animals for 40 individual sites.

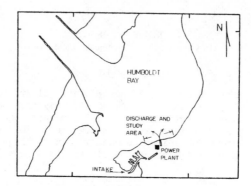

Fig. 1. Humboldt Bay. Arrows indicate water flow
and the length of shoreline investigated.

Temperature

Adams and his co-workers measured surface water temperatures at each of the transects throughout the tidal cycle. For each set of measurements, they used the lowest temperature as ambient and the increase for the other transects was taken as the increment attributable to the power plant. Average temperature increments were calculated from at least twenty measurements at each transect.

Calculation of Similarity Coefficients

The analysis assumed that similarities between the biological communities at each site could be described numerically and related to abiotic factors. Sorensen (1948) first described the particular similarity coefficient used in this analysis: S_{ij}, the similarity between communities i and j, was the number of species the communities have in common divided by the average number present. The similarity coefficient thus ranged from 1.0 for communities with identical species lists to zero for communities with no species in common. The 40 algae species lists and 40 animal species lists generated two matrices of 780 unique similarity coefficients. An initial comparison showed that the algal and animal similarities for each pair of sites were not related in a simple manner. The two sets of similarities were therefore analyzed separately.

Exploratory Principle Components Analysis

Analysis of the data began with resolution of the matrices of similarity coefficients onto principle components. The methods represented the similarities as linear functions of 40 orthogonal component vectors.

This exploratory analysis for both matrices indicated:

1. Two components reproduced a major part of the similarity coefficient variance.

2. No other individual component represented a significant amount of the variance.

3. The two significant components were correlated with tidal levels and average temperature increments at the sites. Temperature was less

important for animal coefficients than plant coefficients, and had
little effect on either for sites at the highest tidal level (1.5-2.1 m).

Least-Squares Best Fit Analysis

Because numerical values for tidal level and temperature were both available, the
analysis was completed with a direct least-squares best fit between the similarities and these two parameters. The success of the principle components analysis
suggested a like functional form for the best fit approximate:

$$S_{ij} \approx K - \sqrt{(K_L \, \Delta L_{ij})^2 + (K_T \, \Delta T_{ij})^2}$$

where ΔL_{ij} was the difference in average tidal level between sites i and j, ΔT_{ij}
was the difference in average temperature increment between sites i and j, and K,
K_L and K_T were calculated by the least-squares best fit program. (The lack of
temperature effect at the highest tidal level was accounted for in the calculation
by setting the temperature increment to zero for all of those sites.)

The approximation corresponds to an ordination of the sites as points in two-
dimensional space such that the axes of the space represent tidal level and temperature and the distances between the points represent the differences between the
sites, $K - S_{ij}$. K represents the similarity between two sites at the same tidal
level and average temperature. The value of K is less than 1.0 because the survey
is not perfect, and because the biological communities are subject to non-systema-
tic variability. Where no difference in average temperature exists, K_L is the
coefficient for reduction in similarity because of tidal level differences. For
two sites at the same tidal level, K_T is the coefficient for reduction in similarity because of average temperature increment differences.

Results

The least-squares best fit produced values and error estimates for the three constants (Table 1). The correlation coefficients between measured and fitted values

TABLE 1.

Results of least-squares best fit ordination for
Humboldt Bay biological communities*

	K	95% Confidence Limits	K_L (m^{-1})	95% Confidence Limits	K_T (C^{o-1})	95% Confidence Limits
Plant Communi- ties	.66	.63 .68	.20	.17 .22	.064	.046 .083
Animal Communi- ties	.73	.71 .75	.15	.13 .17	.015	-.009 .040

*
$$S_{ij} \approx K - \sqrt{(K_L \Delta L_{ij})^2 + (K_T \Delta T_{ij})^2} \, .$$

for the similarity coefficients were 0.60 and 0.54, suggesting the approximation was
effective. The corresponding ordination of the sites was prepared by plotting K_L
times tidal level versus K_T times average power plant temperature increment for

234

each site (Fig. 2).

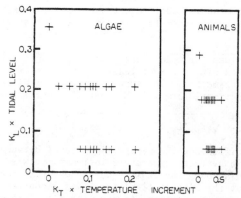

Fig. 2. Ordination of intertidal biological communities at the
Humboldt Bay power Plant. Plotting the weighted values of
tidal level and temperature for each site produces inter-
point distances correlated with $K-S_{ij}$. The upper point in
both cases represents all fourteen high intertidal sites;
temperature effects are negligible. Some other points also
represent more than one site. The greater effects of
temperature on the algae are apparent.

Discussion

Similarities among plant communities and similarities among animal communities did
not vary in the same way within the study area. Regression of the animal similarity coefficients for each pair of sites against the corresponding plant similarity
coefficients produced a best fit line with a correlation coefficient of 0.48 and
an intercept of 0.40. Thus the relationship was only moderately consistent, and
some pairs of sites with quite different algal communities supported animal communities with a significant number of species in common.

The difference results primarily from the lesser effect of temperature on the animals. K_T for the animals was only .02 per C^o, and the 95% confidence limits included a value of zero. The algal similarities declined by .064 per C^o, a value
significantly different from zero. No explanation for this difference is provided
by the analysis done here, but a previous similar observation has been made. North
(1969) suggested that algae in the area of the discharge from the Morro Bay power
plant were more seriously affected than the animals.

The algae coefficient of 0.064 per C^o is smaller than the 0.10 per C^o reported in
previous work (Devinny, 1978). The Humboldt Bay survey sites were much smaller
and closer together, and some may not have been entirely independent. Either of
these values represents a significant biological effect. A similarity reduction of
.064 per C^o means community species composition changes by 6.4% for each degree of
temperature change. The species lost may have significant ecological, aesthetic,
or even economic value.

The analysis did not rule out effects from factors which might be closely correlated with temperature. Neither chlorine nor other biocides are used at the plant,
however, and the reasonable agreement with previous results for other outfalls and
natural temperature gradients suggests no additional factors need be considered to
explain the results.

The elevated temperatures had less effect on the high intertidal organisms, which are exposed to the atmosphere for a large fraction of the tidal cycle. Air temperature varies much more than water temperature, and sunlight may further heat the organisms while they are exposed. Thus the higher intertidal organisms are limited by the more harsh terrestrial environment, and population distributions are little affected by increases in water temperature.

The change of the biological community with tidal level is widely recognized. A previous analysis (Devinny, 1978) showed algal community similarities reduced by depth differences. Communities changed with depth more rapidly in the shallows. Similarity reductions ranged from 0.048 m^{-1} for the interval from 0 to 3 m deep to 0.003 m^{-1} for 3-meter intervals 10 to 20 m deep. The value found here for the intertidal sites, 0.20 m^{-1}, suggests elevation has considerably stronger effects in intertidal zones. This is consistent with the generally accepted observation that periods of exposure to air are a powerful controlling factor for intertidal algae.

Acknowledgements

The results obtained here were critically dependent on the excellent data collected by Adams and his coworkers. Albert Change, of the California Institute of Technology Computer Center wrote the program used for the least-squares best fit analysis. Art Jensen's program was used for correlation calculations. The work was performed while the author was a graduate student under the direction of Wheeler North. Claudia McMahon aided in manuscript preparation.

References

1. Adams, J.R., 1975. "The Influence of Thermal Discharges on the Distribution of Macroflora and Fauna at Humboldt Bay Nuclear Power Plant, California," Ph.D. Thesis, University of Washington, Seattle, Washington.

2. Devinny, J.S., 1978. "Ordination of Seaweed Communities: Environmental Gradients at Punta Banda, Mexico," Bot. Mar., 21:357-363.

3. North, W.J., 1969. "Biological Effects of a Heated Water Discharge at Morro Bay, California," Proc. VIth Int. Seaweed Symp., 275-286.

4. Pacific Gas and Electric Company, 1973. "An Evaluation of the Effect of Cooling Water Discharges on the Beneficial Uses of Receiving Waters at Humboldt Power Plant," San Francisco, Ca., 250 pp.

5. Sorensen, T., 1948. "A Method of Establishing Groups of Equal Amplitude in Plant Sociology Based on Similarity of Species Content," Kgl. Dan. Vidensk. Selsk. Biol. Skr., 5:1-34.

INDEX

Acartia tonsa 77, 84, 85
Accelerogram 11, 12
Accurate Delineation 16
A. D. Little Inc. (ADL) 188, 190, 191
Alcan Plan 95
Alyeska
  oil pipeline 95-99
Anaerobic
  resuspending sediments 64
  conditions 58
Anchovy 67, 87
Anoxic
  waters 63
  conditions 64
Anthracene 129
Arctic Gas Plan 95-97
Aroclor 1254 115-117
Asphaltene 58

Baldwin Hills Oil Field 26
Beta-Sitosterol 134, 139
Biocides 227
Biota 66, 67
Bitumen 52, 55, 56, 58, 59
Biological Oxygen Demand (BOD) 66, 67

California Energy Commission (CEC) 190
California Environmental Quality Act
  (CEQA) 220
Carbopack A. 125
Carbowax 1500 125
Carcinogens 121, 129
Chlorinated Water
  hydrocarbon pesticides 114, 115
  naphthalenes 114, 115
Chlorophyll 77
Cladding 142, 146
Cladocerans 77
Clean Air Act 159
Continental Shelf 3, 6, 37, 58, 197
  201, 207, 208
  North Sea Continental Shelf Cases
  198
  Outer Continental Shelf Lands Act
  199-205
Crustaceans 67, 88
Cyclopoid Copepod 77
Delayed Mortality 104
Diatoms 77

Dichlorobenzenes 121
Dieldrin 115

Earthquakes
  Ground Motion 10-12
  Faults 12-14
  Seismicity 14, 16-20
  Frequency Magnitude Statistics 20-24
  Monitoring 24-27
Echinoderms 88
Ecosystem 88, 224
El Paso Plans 95
Environmental Impact Report (EIR) 187,
  188, 190, 191
Estuarine System 63
Evadne nordmanni 77
Exoskeletons 88

Federal Energy Regulatory Commission
  (FERC) 187, 194
Federal Power Commission (FPC) 187

Gas Chromatography (GC) 114-140
Gastrapods 104
Gaussian Solution 190

Halogenated Compounds 121
Halowax 1014 115
Heterogenity 103
Heptachloroviphenyl (m/e 358) 115
Holocene 12
Hydrocarbon 53, 54, 57, 121, 128, 200
  in-water monitoring 142, 146, 151,
  155, 156

Ichthyofauna 87
Invertebrates 62
Izonization Detector 114, 125

Kerogen 52, 53, 55-59
Kinematic Modeling 188

Liquefaction 14, 142
Liquefied Natural Gas (LNG) 181-194
  benefits 182
  costs 182
  risks 181, 185-188, 191, 192
  terminal EIR 188
  spills consequence 188, 189